ENGINE DETAILING

GW00480600

David H. Jacobs, Jr.

Motorbooks International
Publishers & Wholesalers ®

First published in 1992 by Motorbooks International Publishers & Wholesalers, PO Box 2, 729 Prospect Avenue, Osceola, WI 54020 USA

Motorbooks International books are also available at discounts in bulk quantity for industrial or sales-promotional use. For details write to Special Sales Manager at the Publisher's address

Library of Congress Cataloging-in-Publication Data
Jacobs, David H.
    Engine detailing / David H. Jacobs.
        p.   cm.
    Includes index.
    ISBN 0-87938-610-X
    1. Automobiles—Motors—Cleaning.   2. Automobile detailing.   I. Title.
    TL210.J25   1992
    629.25′04—dc20   92-9009

On the front cover: A concours-winning 1959 Jaguar XK 150 S shines in the sun while its engine undergoes some touch-up detailing. Its owner is John Hall of Mercer Island, Washington. *David H. Jacobs, Jr.*

Printed and bound in the United States of America

Contents

Acknowledgments

Engine and engine compartment detailing encompasses a wide array of enthusiastic and innovative rejuvenation and restoration activities. A street rodder may view this subject a bit differently than a Concours d'Elegance competitor, as would a professional automobile restorer's insight vary from that of a muscle car enthusiast. With such a diverse audience, I felt it was important to present a number of perspectives on the subject. To that end a lot of folks offered their two cents' worth and many more volunteered their automobiles as guinea pigs, hoping to get a free engine detail out of it. But it was the following enthusiasts and professionals who really pitched in and helped, and I want to thank them now.

Dan Mycon, owner of Newlook Autobody in Kirkland, Washington, is a patient and enthusiastic professional and avid street rod enthusiast. I appreciate the time he spent with me to explain a rodder's approach to engine detailing, and the hours he let me hang around his shop to photograph some interesting automobiles, including his 1948 Chevrolet that was currently being transformed from a hulk into a spiffy street rod.

Long-time auto enthusiast and car connoisseur Art Wentworth of Bothell, Washington, had plenty to say on the subject, as usual, and I appreciate him sharing those thoughts with me. I also want to thank him for posing in a number of the photographs and for completing several engine detailing tasks in front of the camera. Jerry McKee of Seattle offered use of his concours-winning 1972 Jaguar XKE V-12 Roadster for photo sessions and was delighted when Wentworth volunteered to spend a few days helping to detail the engine compartment in preparation for an upcoming local concours event.

Thanks also to George Ridderbusch of Everett, Washington, who has won just about every Concours d'Elegance award possible with his 1979 Porsche 928. His insight into the sport is second to none and I appreciate the time he spent with me. The same goes for Dan Case, of Bellevue, Washington, Jaguar aficionado and concours judge. He offered not only a judge's perspective, but a contestant's as well. Thanks also goes to John Hall of Mercer Island, Washington, for pulling his prized 1959 Jaguar XK 150 S out of its trailer so pictures could be taken of its pristine engine compartment. I appreciate the time and effort he has put into this perfectly maintained automobile.

I want to thank Jim Poluch and Christine Collins, executives from The Eastwood Company of Malvern, Pennsylvania, for their support. They supplied lots of information, samples, and photographs of special tools and equipment frequently used by serious auto restorers and detailing enthusiasts. Appreciation is also noted for the support and cooperation received from John Pfanstiehl, president of Pro Motorcar Products, Incorporated, of Clearwater, Florida. He was gracious enough to send photos of factory-original parts from his absolutely original 1959 Cadillac Eldorado, as well as samples of his company's automotive products.

Camee Edelbrock, director of advertising for Edelbrock Automotive Performance Products of Torrance, California, supplied a number of photos featuring Edelbrock's special automotive products and I thank her for her support. A hearty thank you goes also to Larry and Karen Johnson of Corvette & High Performance, Incorporated, of Olympia, Washington, for making sure I got into their Corvette and High Performance Meet and allowing me plenty of time to shoot photographs. I also extend my appreciation to RB's Obsolete Automotive, Incorporated, and Super Shops, Incorporated, both of Lynnwood, Washington, for letting me photograph some of their special automotive accessories.

I am grateful for Terry Skiple's views on engine compartment detailing from a dealer's standpoint, and also for making his 1974 Pantera available for a photo session. He lives in Seattle and is the manager of a large Chevrolet dealership in Kent, Washington. Janna Jacobs' 1988 Volkswagen Jetta and Shannon Turk's 1976 Toyota pickup were used as detailing models and I want to thank them for providing cars in such dire need of an engine compartment detail. I think they both got the best end of the deal, though!

It is imperative that I thank Van and Kim Nordquist of Photographic Designs in Everett, Washington, for doing such an outstanding job processing film and hundreds of prints to perfection, taking extra time to make certain each one turned out just right. And thanks to freelance editor Sharon O'Donnell for her fine work on the original manuscript.

Finally, I want to extend my appreciation for the continued support and encouragement offered by Tim Parker, Barbara Harold, Michael Dregni, Greg Field, Michael Dapper, and Mary LaBarre of Motorbooks International. Their professionalism and editorial assistance has helped to make this a rewarding book project.

Introduction

The overall fascination, interest, and genuine appreciation for automobiles shared by millions of avid enthusiasts throughout the world becomes somewhat complex and varied when one realizes the many different ways to "fix up" a car or truck today.

One person might prefer lots of chrome, high-gloss, and custom goodies while another may opt for simplicity and factory originality, and pay strict attention to factory shop manual specifications. Everyone, it seems, has his or her own idea of what the perfect car or truck should look like. A street rodder, for example, may prefer a big V-8 with a supercharger poking out of the hood on a 1956 Chevy, whereas a Concours d'Elegance competitor with an identical make and model would be most concerned about maintaining total originality and preserving or rejuvenating each and every stock part, with no interest in aftermarket parts or custom add-ons.

This specialization of automotive interests is not limited to paint jobs, wheels, or interiors. It carries over to engines and engine compartments, as well. Just look through any assortment of auto-related magazines and you will quickly notice that some are geared toward custom hot rod engine modifications, others depict ways to perfectly restore mechanical parts to original factory condition, and a few may concentrate on high-performance parts and racing concerns.

What this all boils down to is that engine and engine compartment detailing may mean different things to different people. From a simple perspective, engine detailing can be broken down into three basic categories: (1) A general enthusiast's efforts to clean up an older regular driver with a high-pressure washer and cans of engine paint, gloss black, and clear lacquer; (2) a street rodder's work to build an eye-catching machine with lots of chrome or billet add-ons and plenty of custom parts and accessories; (3) a concours competitor's labor-intensive effort to straighten each fin on a radiator and clean every nook and cranny on the engine and in the engine compartment, even if it means using cotton swabs to reach into the tiny cracks and crevices.

As impossible as it may seem to get two car enthusiasts to agree on which wax or polish works best, the same confusion exists over what engine detailing really means. Mention engine detailing to one enthusiast and he or she may conjure up thoughts of bright neon ignition wires and braided radiator hoses. To an antique auto restorer, it might inadvertently remind him or her that a tiny spot of rust in a paint chip on a motor mount needs to be repaired before an upcoming concours competition.

For all intents and purposes, engine and engine compartment detailing must initially focus around general washing and then graduate to in-depth cleaning—you can't detail a dirty engine. Before you begin thinking about custom add-ons, show car status, or concours trophies, your car or truck's engine and engine compartment must be cleaned from top to bottom, side to side, and front to back.

As varied as definitions for engine detailing may be among auto enthusiasts, so are those for engine cleaning. An inexperienced detailer might think that a simple rinse with a self-serve pressure washer and a few swipes with a cleaning cloth are sufficient, whereas a serious auto aficionado would never consider the use of high water pressure and will expect to spend at least a day or two gently cleaning the engine and engine compartment with as little water as possible and as much attention to detail as can be mustered.

Most serious do-it-yourself enthusiasts prefer to use an old-fashioned and time-tested method for engine cleaning, called elbow grease. Although a few novice engine detailers have gotten away with a quick cleaning by using chemical wheel cleaners to brighten some aluminum engine parts, conscientious car owners most commonly rely on gentle engine cleaning with soft cloths, paintbrushes, toothbrushes, cotton swabs, mild degreasers, and the like. Their attention is focused upon thorough cleaning, of course, but consideration is also given to long-term preservation of all engine compartment surfaces. Using a chemical wheel cleaner will surely brighten some unpainted parts, but what are the long-term effects on electrical connections, belts, hoses, and other parts exposed to such potent chemicals?

This book enlists engine detailing tips, techniques, and advice from avid auto enthusiasts, successful Concours d'Elegance competitors, car show judges, professional autobody specialists, and product manufacturers. Although their methods may differ slightly from those of other successful engine detailers, they have been proven both effective and safe. However, if you should find a simpler and quicker way of accomplishing a particular detailing chore, then by all means use it. The bottom line is overall engine and engine compartment cleanliness and part preservation.

A special automobile with an excellent paint job, detailed exterior, and good-looking wheels and tires may appear to be a thing of beauty as it cruises down the street, but it can only be truly admired when it is parked. Raise the hood of that car to show off a perfectly detailed engine compartment and you will impress even the most uninterested passerby.

As much as a finely detailed engine can enhance the overall appearance of almost any car or truck, a filthy, neglected, or haphazardly detailed engine compartment can surely detract from even the most elegant vehicle. Take your time, follow the advice and suggestions offered on the following pages, and have a good time detailing your favorite automobile or truck's engine and engine compartment.

You might not achieve perfection the first time around, but continued practice and attention to detail should improve your ability to easily and quickly spot minor flaws or oversights and correct them. Soon enough, with patience and persistence, your keen eye and engine detailing expertise might someday qualify you as a Concours d'Elegance judge and owner of a concours-winning automobile.

Chapter 1

Why Detail an Engine Compartment?

To a casual observer or day-to-day automobile operator, thoughts of detailing an engine compartment might seem unnecessary or downright repulsive. Sure, everybody likes to drive a clean car, but who really cares about having a clean engine? After all, isn't that why mechanics always have plenty of rags around their shops?

Well, as any avid auto enthusiast knows, a clean engine and engine compartment offers a lot more than just something to brag about or show off. Removal of grease and grime deposits from under a hood allows linkages and other moving parts to operate freely and easily. An engine runs cooler

without blankets of grease, and it is a lot simpler to perform routine maintenance on a clean engine than it is on one that looks like a big sludge ball is jammed into its compartment space.

Some people attempt to detail their car's engine once in awhile, even though they may not know it. The owner of a brand-new Cadillac, for example, might wipe off an air cleaner, battery, and hood underside periodically, just because he or she likes the new car and wants to keep it looking good. Such minimal efforts cannot be construed as all-out engine detailing by any means, but they are well intended, nonetheless. Unfortunately, however, those small ex-

The engine and engine compartment on this 1976 Toyota pickup have been neglected for far too long. The obvious build-up of grease and grime will require major degreas- *ing and cleaning in order to make things look presentable. Along with its unkind appearance, mechanical problems could lurk in there, too.*

ercises soon become futile as grease and road grime eventually cover much more than just an air cleaner or hood underside.

Once an engine and engine compartment have been meticulously detailed, minimal systematic efforts are all it takes to keep them looking good. Each time you wash your car, for example, take a few minutes to wipe off accessible engine compartment areas, parts, and accessories. The benefits you'll gain by maintaining a clean and detailed engine space will far outweigh any amount of labor that goes into the endeavor. Since grease and oil accumulations always attract more and more debris to eventually establish large build-ups of grit and grime, you will be way ahead of the game by cleaning small splotches of grease or oil early on, before they become major cleaning dilemmas.

Auto enthusiasts detail engine compartments for lots of other reasons. Some detail fanatics, like Art Wentworth, simply can't stand the thought of any part of their car or truck being dirty. They insist that everything look, feel, and smell clean. Others like to tinker with engines and perish the thought of having to tune up a filthy motor. For show car and Concours d'Elegance competitors, clean engines and engine compartments are mandatory.

Engine compartment detailing is not difficult to learn or accomplish, nor will you need a lot of special tools or equipment. If you are as meticulous about your car as Art Wentworth, then in-depth engine detailing will be somewhat time consuming and probably require lots of elbow grease; there are many small, hard-to-reach areas inside the engine compartment for dirt, grease, grime, and debris to accumulate. For most of us, time and elbow grease are easy to come by. And, once your major engine detail is accomplished, routine maintenance will keep it and surrounding compartment areas looking great for a long time.

Pride in Appearance

Have you ever seen a favorite vehicle make or model at a car show that had a great-looking interior and exterior but displayed an ugly engine compartment? Maybe one with dull or chipped paint, flaking chrome, antifreeze stains, or small pockets of light-colored dirt around the entire compartment? If so, how did the condition of that engine compartment

The nitrous oxide system on this pristine 1984 Corvette will certainly help make the car go fast, and the beautiful condition of the engine and engine compartment is certainly something to admire. Maintaining the engine compartment helps an owner to quickly spot mechanical problems and also helps to promote maximum performance from all engine parts and accessories.

Clean linkage assemblies, cables, and other shaft-mounted revolving engine parts always operate best when free of dirt and grime build-up. It is much easier to notice frayed cables or loose connections on parts like this when they are kept clean.

affect your overall opinion of the vehicle? Were you disappointed?

Engines are typically viewed as focal points for most automobiles. Just about every car show will feature entries with hoods up, and used car dealers frequently leave hoods open during working hours to show customers how completely their cars have been cleaned, maintained, and generally cared for. For most people, a clean engine and engine compartment project an impression that the owner cares about the vehicle because he or she took time to clean and maintain everything under the hood. First impressions are powerful, and clean engines almost always make positive statements about vehicles and their owners.

From a street rodder's point of view, an engine might only look its best when displaying lots of chrome or billet goodies. From the standpoint of someone who prefers stock machines, a super-clean and original factory-painted engine may prove most noteworthy. In both cases, however, accumulations of oil, dirt, or other debris is destructive to the highest degree, just like that favorite show car with a great

body and interior but a mediocre or poorly maintained engine compartment.

Auto enthusiast and concours competitor Jerry McKee bought his 1972 Jaguar XKE Roadster new. The two of them have traveled many miles together. Even though the car is frequently driven on fair-weather days, it still looks great. In fact, it recently won first-place honors in its category at a local concours event. Although he is not as detail oriented as Art Wentworth, McKee has a lot of pride in his automobile and has always maintained it in a serious manner, including keeping the engine compartment clean, polished, and original.

When McKee unlatches and tilts the front end of his E-Type forward to expose its long V-12 engine, one quickly notices the lack of any road grime or engine lubricant build-up. The engine may not be perfectly detailed, since it is frequently driven and has never been restored, but efforts to keep grease, oil, and dirt accumulations to a minimum are more than obvious. The car's clean and maintained engine compartment enhances its overall beauty and certainly reflects McKee's pride of ownership.

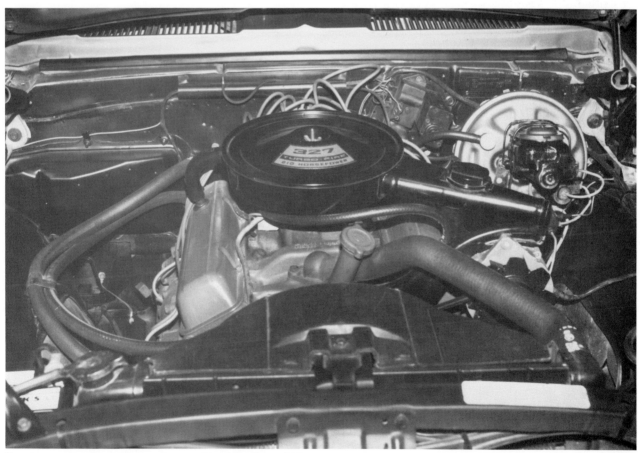

Routine preventive maintenance on this tidy Chevy 327 V-8 are a snap when compared to identical chores on a sludge-covered, neglected engine. There are no globs of grease in the way, or worries of debris falling onto engine parts.

Improved Performance

People generally feel that their cars run better after they have been washed, had the windshield cleaned, and had all the sticky substances wiped off of the steering wheel. Although no mechanical parts had been touched, folks actually believe they experience much better vehicle performance after a car wash. It may be that they just simply feel better driving a clean car.

Engine detailing, on the other hand, should realistically improve mechanical performance. Not only will linkages and cables operate smoother with less resistance, but an engine should also run cooler because its clean surfaces are now able to disperse heat more readily. Along with that, clean electrical connections are better equipped to conduct ample current flows, and the alternator and air conditioning compressor should function more efficiently with accumulations of dirt and crud removed from vent slots and around operating shafts.

Typically, conscientious do-it-yourself engine detailers should repair minor mechanical problems as they are recognized during detail chores. Maintenance corrections such as tightening loose bolts, nuts, and screws can go a long way toward improving overall vehicle operation. Going a few steps further to replace hoses suffering cracked ends, frayed or notched belts, rusty electrical connectors, soft radiator hoses, dirty filters, and the like should increase fuel economy and help an engine run strong and smooth. A clean engine might even entice you to perform an in-depth tune-up, with quality parts and meticulously accurate adjustments, to bring it up to peak performance.

Small oil drips are easily detected on a clean engine. Once located, steps taken to correct leak problems will not only help keep engines clean, but also assist inadequately tightened, sealed, or adjusted parts to operate more effectively. Prolonged oil leaks

Obviously, the person who attempted to wipe down the underside of this hood did not take the task seriously. Such futile efforts will not do much to improve the appearance of engine compartments, nor will they help limit amounts of dust and dirt the remaining grease and dirt films will attract.

The engine and engine compartment on this restored 1958 Chevy Impala is stunning. The engine is painted the same shade of red as the inner fender panels and body parts. Chrome strips are featured on valve covers and the fuel-injection assembly. You can bet that the owner of this machine takes time each week and after each drive to wipe down the engine compartment.

on hoses and wires will soften their coatings and speed up deterioration. Lower radiator hoses softened too much could easily collapse under high engine speeds to wipe out a cooling system and possibly the engine itself.

This type of routine engine maintenance during or right after engine detailing should also be considered for accessory parts located throughout an engine compartment. Many times, minor operational part failures such as faulty windshield washers, horns, and cruise control units are simply caused by loose wires, bad connections, split hoses, or frayed cables. Loose or missing shrouds, for example, could cause a definite reduction in a cooling system's efficiency and cause a motor to run hotter than designed—something which could quickly lead to major engine problems.

Covered with a thick coat of grease and crud, these kinds of minor problems easily go unnoticed and possibly lead one to believe that an entire unit is broken, which will cost a great deal to replace. However, once parts and adjoining connections are cleaned and made clearly visible, one might discover that repairs will only require a few minutes' time and a couple of dollars' worth of related parts. Maintaining a clean and operationally efficient engine will not only improve performance, but should also go a long way toward helping the engine and its related parts last a long time.

Equipment Longevity

You wouldn't expect your car's paint job to last for ten years if it was never washed or waxed and always parked in bright sunlight next to the ocean and its heavy, salt-laden air. Nor could you expect the tires on your car to last long if they were never rotated or never had their air pressure checked and then maintained at manufacturer specifications. The same theory holds true for everything in your car's engine compartment.

Engine and engine compartment of a concours-class 1954 Jaguar XK 120 M. Note how clean the area is in front of the radiator, and how the featured wires are neatly bundled. All polished aluminum on the engine shines brightly and there is not a speck of dirt or dust anywhere inside the compartment. Cars have to look like this in order to be serious concours contenders.

This poorly maintained 396 V-8 detracts from the overall appearance of its host 1968 Chevrolet Camaro. The block is in dire need of fresh paint, and everything else could use sprucing up, too. Neglected engine compartments can quickly destroy an admirer's perception of any vehicle, even if the interior and exterior are perfect.

11

A radiator cannot be expected to lower engine coolant temperatures if its fins are caked with old bugs, pebbles, and road grime. Sooner or later, a neglected radiator fails and the engine overheats to blow the head gasket, crack, or just seize up. Other potential problems lurk under the hoods of neglected engines too, like failing power steering units, water pumps, and alternators with belts that are frayed and notched or supporting bolts that are so loose that the units wobble with every engine revolution.

Very high temperatures are common inside an engine compartment. Although automotive engine accessories are designed to withstand certain temperature ranges, years and years of heating up and cooling down will take its toll on neglected parts. Cleaning and detailing engine compartment parts helps them to last longer by ensuring adequate protection with recommended coats of paint, protectant, or lubricant. Ignition wires might last longer when they are correctly routed away from hot ex-

haust manifolds and secured in position by supports or brackets. Wires, hoses, and fuel lines could suffer chafing damage if not held in place with properly mounted, sturdy connectors. In addition, these items always look their best when routed evenly and uniformly.

Minor problems are easy to detect on detailed engines. Should a coolant, fuel, or other fluid leak appear, you should be able to quickly recognize and repair problems early on. Conversely, those same potential problems on dirty engines might go unnoticed for a long time, so long, that you may not know there are problems until the engine overheats, fails, or catches on fire. When engine problems get that severe, it is easy to understand how maintaining a detailed engine compartment can help promote extended operational longevity.

Overall Value

Having been in the car business for more than 20 years, Terry Skiple is an experienced new and used

Engine compartments are commonly regarded as focal points for many automobiles, especially those featured in car shows. A lot of effort has gone into the serious detailing of this fifties vintage Chevy pickup truck engine compartment, as noted by its sanitary condition and pinstriping accents.

Displayed on the sales floor of RB's Obsolete Automotive, this engine should make any street rodder smile. The billet accessories and braided hoses look good, and the perfectly clean condition of the entire unit is awesome. Once your engine is clean and tidy, the installation of parts like these should make you smile, too.

automobile sales manager. He knows that interior and exterior automotive detailing will bring in more money for almost any used car. Detail an engine, too, and the asking price goes even higher. This is why almost all used car businesses have trade-ins promptly detailed before they are displayed on the lot. Seldom will you find used cars on a dealer's lot that have not undergone complete details, including engine compartments.

The same detail philosophy Skiple applies to his business holds true for you as a consumer as well. If you want the highest trade-in allowance for your used car, detail it before bringing it in for a trade-in appraisal. Be sure to open the hood and make mention of the engine compartment's excellent condition and well-maintained environment. Pristine automobiles almost always bring in top dollar, and yours should be no different.

From a dealer's standpoint, a well cared for and properly detailed automobile can be put on display

This V-12 engine that powers Jerry McKee's 1972 Jaguar XKE Roadster has never been restored or professionally detailed. However, McKee maintains a systematic routine of wiping it down soon after each drive and never lets oil or grease build up. His pride of ownership is certainly reflected in the well-maintained condition of the engine area.

Removal of grease and crud from around the distributor and the rest of the engine compartment will improve the engine's ability to perform as expected. Thick blankets of grease tend to make engines run a little hotter than normal, reduce the efficiency of moving parts, and interfere with those parts designed to conduct electrical current.

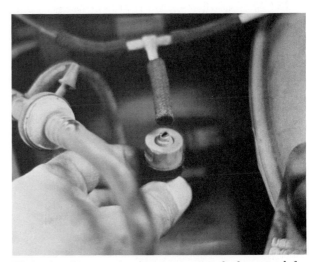

While detailing engine compartments, look around for broken parts, loose connections, and other misaligned items. Make repairs or adjustments as needed, or plan to complete required maintenance right after detail chores are finished. This kind of attention to detail and preventive maintenance should help keep your car or truck's engine in top visual and mechanical condition.

13

right away. He or she does not have to wait for a professional detailer to schedule it for service, nor will there be a detail fee charged against it. This saves a dealer time and money, and also allows salespeople to boast about what an exceptionally well maintained used car it is and how it is worth top dollar. Get the picture?

As much as a dealer may appreciate a detailed trade-in, a prospective buyer for a used vehicle you are trying to sell may be very impressed with a pristine engine compartment and be willing to pay you top dollar. On the other hand, a sloppy engine detail that reeks of multipurpose dressing on hoses and wires and shows excessive engine paint over-spray splotches on unpainted carburetor parts, linkages, and belts, should immediately raise red flags for any knowledgeable buyer and would, therefore, work against you.

These kinds of detailing flaws scream out that an engine compartment has been recently detailed (by a sloppy amateur, no less) and should make one wonder what that area looked like before and what, if any, kind of preventive maintenance schedule had ever been sustained throughout the drivetrain's lifetime.

Equally as important, you should make note of overall engine compartment conditions when considering the purchase of a used automobile. For one like that just described, you may be able to haggle over the purchase price with some authority and achieve a cost reduction. Or, you may opt to pass on that vehicle and continue looking for one that offers more quality by way of its appearance, condition, and operation.

Engine Detailing Concerns for Special Vehicles

Out of the regular used car arena, owners must be keenly aware of what they are doing when detailing engine compartments in the antique, classic, and specialty car segments. This is critically important for owners of stock factory machines, especially if they have ever entertained intentions of entering concours competitions.

Some manufacturers placed identification numbers on almost every engine part installed on their automobiles. These numbers are crucial when attempting to prove originality. Their destruction or disfigurement could be devastating with regard to winning concours or accomplishing maximum restoration value.

The owner of this 1981 Ferrari left no doubt about when its next oil and filter change was due; the mileage and date of its last servicing are written clearly on the filter. It is *always a good idea to service engines in a systematic and timely fashion, and what better opportunity than while performing weekly or monthly detailing tasks?*

The classy appearance of this Chevy V-8 with fuel injection is enough to make one believe it runs strong. It does, and the regular maintenance it receives goes a long way toward achieving that end. Keeping the engine and its compartment assemblies clean and well groomed also promotes the parts' longevity and high-performance operation.

The grommet designed to protect these lines against chafing has come loose from its position. This kind of problem would surely result in damage to the lines, unless corrected early on. Recognizing and repairing minor engine compartment flaws, like this one, are an additional benefit afforded those who detail engine compartments on a regular basis.

A pinhole leak is presented by way of an antifreeze stain on the side of this radiator. A small leak like this is much easier and cheaper to repair than a larger one. As you proceed with engine compartment detail tasks, make note of similar coolant, oil, or fluid leaks so they can be fixed as soon as possible, before they grow into larger problems.

The battery hold-down is missing from the engine compartment on this 1974 Jaguar XJ 12 L. Allowed to sit unsecured, batteries could easily fall over when making sharp turns or quick stops. Part of your engine detailing chores should entail securing all such parts.

Certain parts on older cars were originally manufactured and installed unpainted; others were coated with a kind of paint that produced only a certain amount of gloss. For a concours car, this same degree of factory paint gloss has to be maintained in order to earnestly compete and expect to win. This is why serious concours competitors spend untold numbers of hours poring over old shop manuals and any literature that pertains to their cars in hopes of learning how parts were designed to be installed, maintained, painted, repaired, and so on.

An older stock automobile engine compartment can be cleaned and detailed to perfection. Caution arises when detailers get past initial and in-depth cleaning stages and are ready to shine, paint, and otherwise fix up motors and their related accessories. Sure, Chevrolet Orange high-temperature engine paint might be appropriate for a lot of 283s and 327s, but what color adorned the stock engine for a 1935 Sedan Delivery? To maintain originality, one has to consult an older shop manual or seek advice from a knowledgeable concours judge, car club member, autobody paint and supply store jobber, or professional auto restorer.

Stock vs. Custom

Art Wentworth seems to know at least something about almost every automobile ever made. He

Plenty of degreaser and a pressure washer worked a small miracle in making this Toyota pickup's engine compartment look decent again. Compared to what it looked like before, don't you think a dealer would offer more now for a trade-in allowance?

may not be an authority on any specific make or model, like George Ridderbusch, veteran Porsche concours winner and long-time competitor, but he knows what he likes. "I prefer a super-clean, unchanged engine compartment, with parts painted the colors that God made 'em and without a single, solitary custom aftermarket part anywhere," says Wentworth. "I don't even care if paint is mildly peeling or decals are faded, as long as the engine's clean and nothing has been altered."

Ridderbusch agrees to a certain extent, noting that he made sure new body paint on inner fenders inside the engine compartment of his 1979 Porsche 928 maintained a similar degree of orange peel as that which came directly from the Porsche factory when his car was new. But, everything else under the hood has to be perfect in order to score winning points from a concours judge. This calls for meticulous

straightening of radiator fins, touching up cadmium-plated pieces, replacing stock decals with OEM (Original Equipment Manufacturer) parts, and anything else needed to keep things in perfect, original, and better-than-new condition.

If an older rig is slated to be converted into a street rod, or anything other than a factory-stock original automobile, its engine can be painted any color desired: red, purple, green, black, you name it. That's right, newer urethane paint products can withstand high temperatures experienced during engine operations, and custom car professionals are now able to paint blocks any color under the rainbow.

In fact, a lot of serious custom car owners are painting engine blocks the same color as their street rod bodies with exciting results. In addition, a number of maverick and inventive auto enthusiasts have painted just about everything under their cars' hoods, including radiators, alternators, carburetors, power

A stock six-cylinder engine sitting inside the engine compartment of a 1967 Camaro. Fresh paint on the valve cover, head, and block make it look clean and efficient. Surrounding hoses and compartment surfaces need to be cleaned and polished so they can look equally as nice as the motor. Although an engine might look good, full appreciation and maximum vehicle value is only afforded those units that receive complete engine compartment detailing.

When considering the purchase of a used automobile, note overall conditions of the engine and engine compartment. A chrome valve cover may look good, but one in this condition with its oil filler cap missing and spark plug wire holder hanging loosely may indicate that normal preventive engine maintenance was never much of a priority to the current owner.

steering units, and brackets the same color as their car's body with outstanding results. Those engines, complemented with subtle shades of custom-colored wires and hose covers, look absolutely stunning.

Dan Mycon, owner of Newlook Autobody in Kirkland, Washington, is a street rodder and loves to customize cars and trucks. His 1948 Chevrolet was old, tired, run down, and beat up before he started rodding it out; but it was stock. He could have restored it to like-new condition, but never thought about it because of his avid street rod enthusiasm. To him, this car would be boring with a stock motor, three-on-the-tree, skinny tires, and stock wheels. But, drop in a V-8 and decorate it with a lot of billet accessories, hook up the headers, adjust the tilt steering wheel, drop the automatic into low gear, and he's in hog heaven. Especially in the summertime, when he can turn on aftermarket air conditioning.

Although Wentworth, Ridderbusch, and Mycon may have different tastes in automobiles, they share an equal appreciation for cleanliness and attention to detail. Ridderbusch points out that those parts of

engine and engine compartment detailing which include initial and in-depth cleaning, only bring such endeavors to baseline. After that, specialized detail work that involves painting, accessory rejuvenation, labor-intensive gloss perfection, or installation of custom aftermarket parts will either bring a sense of personal satisfaction to a general auto enthusiast, help win a concours event for a serious competitor, or possibly bring in a car show trophy for an active street rodder.

The debate between stock versus custom is not really a loud one. Stock enthusiasts appreciate originality and the time-consuming efforts required to achieve original perfection—not only to rejuvenate old parts, but also to *find* them. Street rodders and other custom car buffs don't want museum pieces, but would rather have a ride that looks wild, roars loud, and goes fast. To each his or her own. It would be a shame to ruin an otherwise perfectly original vintage automobile by turning it into a hot rod because those originals can never be replaced. But, for one that has long been neglected, you have to

The concours-winning engine of a 1966 Jaguar E-Type Roadster. Note the uniform gloss presented by all polished parts and even the painted air cleaner housing. Even though it looks good, a concours judge could mark off points if paint color shades used on various engine compartment parts are not stock-original. This one, of course, is straight by the book.

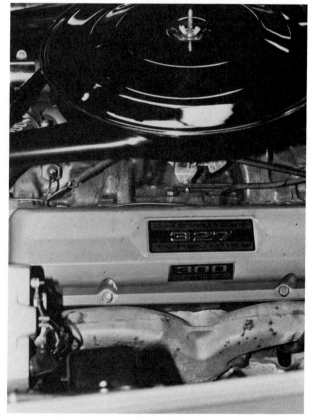

Many avid auto enthusiasts prefer engines and engine compartments that look clean and natural, like this one. However, allowing paint or other surface finishes to deteriorate past the point of no return will cause more harm than good by giving rust and corrosion a place to start. In cases like this, detail efforts must include some paint work.

admit that a pristine street rod or hot rod, put together right with an engine compartment detailed to perfection, looks mighty awesome.

On the other hand, a crisp concours winner with all its hose clamps evenly aligned, hoses that feature identical gloss textures, engine compartment body paint polished to exact standards, and wires routed like a heart surgeon's sutures, looks beautiful.

Along the same lines, a stock muscle car with factory parts, a bit weathered, but clean and original, is a real eye-catcher.

Overview

The choice between absolute concours originality and perfection, street-rod gloss and high-performance aftermarket add-ons, or stock authenticity, lies within you and what you perceive your perfect car to be. It's your vehicle, and you are the one who must complete the work and pay for those parts needed to keep the engine and engine compartment looking crisp and clean.

If you are new to concours or street rodding, or perhaps have just finally found the time and the means to pursue active involvement with muscle cars, race cars, classics, or antiques, and you need some advice or general information, plenty of help is available through enthusiast magazines, car clubs, swap meets, and car shows.

A wide variety of auto-related magazines are published for specific enthusiast interests. These periodicals normally have titles that clearly indicate what kind of information lies between their pages; for example, *Hot Rod* and *The Classic Auto Restorer* are two such publications. Others, like *Skinned Knuckles*, are specialized for serious restorers who look for authoritative articles on the restoration and maintenance of authentic collector vehicles. Along with informative articles, many of these monthly publications feature sections toward the back of each issue that list assorted car clubs around the country. They frequently include the types of vehicles members are most interested in, and addresses where they can be contacted.

Car clubs are usually excellent resources. For the most part, members share an avid interest in certain makes or models or types of automobiles.

The radiator fins and everything else in this area of George Ridderbusch's 1979 Porsche 928 engine compartment look perfect. Notice the uniform and balanced gloss on hoses at the top. This strict attention to detail is mandatory for serious and consistent concours winners.

Some clubs specialize in Corvettes, or Mustangs, or Jaguars, or street rods, or Porsches, and so on. Others are very specific and share interest only in certain years of particular makes and models.

Car club members are generally astute do-it-yourself mechanics, detailers, bodywork technicians, and painters. A few, like George Ridderbusch, have become local experts on certain vintage makes and models. As a car club member, you may be able to get the information needed to accurately restore and detail the engine and all its components on your prized automobile. Not only could members help teach you how to surpass baseline, they may also be valuable assets when it comes time to locate new parts, factory originals, or exact quality replicas.

A valuable aid in helping to locate auto-related literature, car clubs, parts, and service is *Hemmings Motor News*. This telephone-book-sized monthly pub-

Dan Mycon's 1948 Chevrolet undergoes transformation from a neglected heap into a street rod. It could have been restored with its original six-cylinder engine and three-speed tranny, but Mycon opted for a new front suspension, V-8 motor, some custom metal work, and a two-tone custom color. Automobiles can be fixed up in any number of ways, all depending upon the interests and desires of their owners.

The firewall, heater box, fender apron, valve covers, intake manifold, and fuel-injection assembly are all painted the same color red as their host 1958 Chevy Impala's exterior body. The durability of newer catalyzed urethane paint products that can withstand high engine operating temperatures has made this sort of innovation possible. Chrome accessories around the motor accent the red components to make this a very special engine compartment.

Even a die-hard stock-original perfectionist would have to turn around and check out this 1000hp blown 1969 Corvette. Any automobile put together correctly will receive its due admiration from those who understand the time, effort, and money that goes into auto restoration or customizing projects. The key word here is correctly, *with attention to detail along every step of the way.*

lication is a virtual catalog of classified automotive ads that is second to none. In it, you'll find a huge assortment of advertisements for cars, trucks, parts, restoration services, lists of car show events, and a lot more.

Through information in auto-related monthly publications, conversations with concours or car show participants, involvement in a selected car club, or research in shop manuals, you should be able to amass enough information to bring the engine compartment of your favorite automobile up to standards of perfection. Interest in automobiles of all kinds is at an all-time high. With that comes an abundance of pertinent literature and general enthusiast information that is easily obtained by anyone who attempts to find it.

This engine is equipped with Edelbrock Power Package options. An engine like this could be installed in some older vehicles with minimal modification. On others, however, it may require replacement of inferior front suspensions and a lot of custom sheet metal work to make everything fit and operate properly. This is why you must do your homework before tackling major rebuild projects. Photo courtesy Edelbrock Corp.

Hemmings Motor News *is a well-known monthly publication read by thousands of avid automobile enthusiasts. It is jammed with advertisements for all kinds of auto parts, services, and organizations, and can be a valuable guide for anyone contemplating the restoration or rejuvenation of a project vehicle.*

The extra-clean condition of this engine is maintained by conscientious work efforts. The labor and cost involved in building it are another story. Your engine and engine compartment can look just as good, provided you are willing to spend the time and effort and pay close attention to details. Don't be satisfied with second best when you have the ability to do the job right, from the beginning.

Chapter 2

Work Site, Materials, and Tools

As mentioned earlier, the degree to which engine detailing is carried out might be divided into three basic categories: a quick degreasing and simple spiff for an older regular driver in preparation for its sale; a conscientious and in-depth detail for a special, yet

The engine compartment of a regular daily driver. A quick detail to enhance its marketability would include a wash, some in-depth cleaning, and a little black paint work to cover scratches and nicks around the radiator and other engine parts. A detailer for a used car dealer might be inclined to spray clear lacquer over the engine compartment to cover light dirt spots and make it all shine— something nobody should ever do to a nice, well-cared-for automobile.

frequently driven automobile; or the meticulous and labor-intensive perfection required for concours competitions and show car exhibitions. This is not an all-inclusive list, of course, as the intensity of each engine detail ultimately depends on the interest and energy afforded each job by individual owners and detailers.

The appropriate work sites, materials, and tools involved with engine detailing are considerations that must be evaluated before any job is started. Power washing might be a necessity for certain engine compartments, but never a viable option for others. Where a powerful degreasing agent might be best for one project, a mild detergent may be most appropriate for another. You might consider using a wire brush to clean an intake manifold on one engine, but would be compelled to use a very soft toothbrush on another. It all depends on the type of material being worked on, its overall condition, and the kind of results you expect to achieve.

You wouldn't initially power wash a neglected and heavily grease encrusted V-8 in your driveway because of the amount of crud and debris that will most assuredly be left over. However, you might not think twice about working in your driveway when carefully detailing the engine in your well-maintained and pristine family sedan because residue from that job would be minimal, at best.

As you would for any auto-related improvement project, allow yourself ample time to plan a definitive course of action for your engine and engine compartment detail. Envision what you will be doing from start to finish as best you can. Then, using a note pad, make a list of the tools, materials, parts, and supplies you'll need throughout the project so they will be on hand when you are ready to use them.

Work Site Considerations

One would imagine that every automobile enthusiast dreams of having a huge shop with lots of bright lights, electrical outlets, plenty of compressed air capacity and outlets, a hydraulic lift, an assortment of workbenches, power equipment, hand tools, and so on. Big dreamers might also imagine a building that houses a downdraft paint booth and separate detailing wing. Wouldn't it be great to have a large, enclosed stall with hot and cold running water, temperature control, floor drains, a pressure washer, hydraulic lift, and all the amenities close by and set up specifically for automotive detailing?

Unfortunately, few enthusiasts are afforded those luxuries and most of us must settle for more realistic work facilities. On the bright side, though, eager enthusiasts with limited resources can accomplish beautiful engine compartment details in their driveways, carports, garages, or workshops; it may just take them a little longer and require a bit more elbow grease.

Because fundamental cleaning of extra-dirty engine compartments most often requires use of high water pressure and potent solvents, detailers must be concerned about greasy, polluted run-off. Naturally, you will not want this sort of debris to build up on the

deck of your driveway or carport, nor would the environmental agencies want to see it flow into storm drains, sewers, or other drains designed for typical rainwater run-off. Therefore, consider enlisting the services of a facility that is set up for engine steam cleaning or pressure washing and equipped with drain systems that can safely handle such run-off.

Some parts of the country are under more environmental scrutiny than others. Because of this, many auto-related service companies have installed special pollution control systems to safely handle materials they commonly work with or waste they must dispose of. Automotive painters, for example, may be required to install high-tech air filtration systems on top of their paint booths to control overspray pollution, and may also be required to store polluted thinners and reducers in fifty-gallon drums until they are picked up by a hazardous waste disposal company on a weekly or monthly basis.

This 1988 Volkswagen Jetta makes a nice family sedan. Its engine compartment is clean and well taken care of. A conscientious owner could easily justify spending an entire weekend cleaning and servicing this engine and its accessories. Clear lacquer paint should never be used on an engine space that looks this good.

Jerry McKee's E-Type Roadster fits into an engine detailing category just above well-cared-for family sedans. Meticulous engine cleaning and service are a must. Heavy water use while cleaning would be strictly limited to front suspension members only, as water accumulation on this sensitive engine could cause a number of mechanical problems. This special car is worthy of labor-intensive hand cleaning and polishing.

In order to comply with similar pollution requirements, companies that offer engine cleaning services may have had to install drain systems that trap polluted run-off before it enters common sewer lines or storm drains. In those cases, a hazardous waste company might be used to clean sludge from sumps or residue collection traps.

Many self-serve car wash facilities allow engine cleaning in their stalls on a routine basis (always check with an attendant, if available, before starting, to be sure it is permitted). Some may only have one or two stalls designated for engine cleaning, while others may be equipped to handle it in all their stalls at any given time. For a nominal fee, generally $1 to $2, pressure washers can be dialed in to spray clear water or a mixture of water and soap. You will have to supply your own degreasing materials.

One option for getting a first-class degreasing job on both the top and bottom of a heavily encrusted engine and engine compartment is through a professional automotive steam-cleaning outlet. These businesses are generally equipped with hoists or work pits which allow detailers easy access to all areas around engine compartments, as well as entire drivetrains and most vehicle undercarriages. Look under the heading of "Car Washing & Polishing" in the yellow pages of your telephone book for facilities that offer such services. Call more than one to compare prices and available options, as some may not have the means to thoroughly clean lower engine areas or undercarriages.

For lightly soiled engine compartments that do not warrant high-pressure washing and those special ones where only a minimal amount of water will be used, a driveway or drain-equipped ramp should suffice as an adequate work site. A large sheet of plastic placed under the engine area of your car or truck should catch a good deal of solid debris that will be washed off, and also prevent greasy residue

Detailing your engine in a driveway or carport may be much more realistic than hoping for dream facilities. Here, a 1980 Chevy Suburban engine compartment will receive hand washing with minimal garden hose water flow. It is not a greasy engine compartment, so run-off will not be a problem. The fender cover will prevent scratches while the detailer leans over it to reach inside the compartment for cleaning and other work. Fender cover courtesy The Eastwood Company

Art Wentworth prepares to wash down a front suspension section on McKee's E-Type. Water should not be allowed to flow any closer to the engine compartment than this. Its V-12 is very sensitive and has several electrical connections in the valley between valve covers.

from staining the deck. It may be easiest to lay the plastic down first, secure the corners and sides with bricks, rocks, or what have you, and then drive your vehicle on top of it.

Even with a sheet of plastic or large drip pan under an engine, you should be concerned about water run-off. Regardless of the amount of water used, be sure that soapsuds and dirty rinse water will not flow into areas that could be adversely affected, such as flower beds, decorative rock landscapes, lawn areas, and so on. Ideally, water should run away from your work site and into a gutter or drain. Once all engine compartment cleaning chores have been completed, the vehicle should be moved so the detailer can pick up and discard plastic or cardboard, clean off drip pans, and then rinse the driveway or carport area.

After an engine compartment has been initially washed and degreased, in-depth cleaning operations could be conducted in your garage, carport, or driveway. This intricate work should not require much use of water sprays, but will involve working with hand towels and cleaners. Place a large drip pan or large piece of cardboard under the engine to catch small globs or drips of dirt, crud, or cleaner residue that may fall away as detailing progresses.

Plenty of quality light is essential to any detailing workplace. This can be provided through natural sunlight, ceiling- or wall-mounted shop lights, drop lights, or any assortment thereof. Working under the hood of many cars frequently finds detailers straining to see areas located below exhaust manifolds. Shadows cast by hood structures and other obstructions makes intricate detailing work difficult. If you are confronted with these conditions, rig a set of drop lights to fully illuminate your work area so you can clearly see what needs to be done and what progress is being made.

Once an engine compartment has been *thoroughly* cleaned, many detailers believe they have reached a halfway point. This is probably an accurate assessment for those compartments that were already in good shape but might require some minor paint work, polishing, waxing, and general detailing.

Many self-serve car wash facilities allow engine cleaning in their service bays. This is an ideal spot for degreasing heavily encrusted engines, like the one on this 1976 Toyota pickup. You will have to bring along your own degreaser and assorted brushes and cleaning cloths, as water pressure alone is not enough to remove most accumulations of baked-on grease and grime.

This is not true, however, for those engine compartments being prepped for shows, concours events, or custom modifications. Work involving meticulous and intricate detailing is frequently very time consuming and might easily require two or three times as many hours (or more) as those expended during cleaning.

If your car or truck is slated for extensive custom engine compartment modifications or labor-intensive car-show preparations, you should have a workplace available where the vehicle and its related parts can be safely stored overnight or for however long the work may take. This is critically important during inclement weather and for regions located close to saltwater, like coastal shorelines. Moisture that settles on unpainted metal parts can immediately begin a rusting process that will have to be remedied before any paint work is started.

A safe, systematic, and controlled means of parts storage is a major concern for those who plan to rebuild engines and restore engine compartments, and should be resolved long before work starts. Although it may be fun and quite easy to dismantle an engine compartment, putting it back together correctly will be frustrating if parts have been scattered all over or just haphazardly tossed into a big box.

Arrange a set of shelves along an empty shop wall or clear out a particular corner in your garage and designate that space for just your project vehicle's parts. Use a number of small, sturdy boxes as opposed to a couple of big ones. Section out your engine compartment and put parts from individual sections into particular boxes; for example, one box may hold the driver's-side radiator support area, another the passenger's-side inner fender area, another the windshield washer parts, and so on. Heavy-duty freezer bags work great for small items and assorted nuts, bolts, washers, and other fasteners.

Be sure to clearly label boxes with a dark-colored marker. In addition, you may want to include

Wentworth uses Simple Green and a soft paintbrush to attack dirt along the hood underside of McKee's E-Type. Dirt residue will be wiped up with an absorbent cloth or towel. Mild cleaning endeavors such as this can easily and safely be accomplished in driveways, carports, and garages, since there will be very little water flow and no need for high water pressure power.

On most vehicles, working under the hood presents detailers with a serious lack of sufficient light. Illuminate those spaces with drop lights or portable spotlights. This do-it-yourself model features an extra-bright halogen outdoor light fixture with an attached four-plug electrical socket. A set screw on its welded collar allows the light to be adjusted up or down on a support pole.

John Hall uses a wide artist's paintbrush to dust off part of the engine in his pristine 1959 Jaguar XK 150 S. When restoration started on this car a few years ago, simple cleaning didn't even come close to reaching the halfway mark. Now, when he prepares it for a concours event, *about a third of his time is spent cleaning, a third goes into polishing, and another third is dedicated to smaller tasks like touching up tiny paint blemishes, removing dry polish, and dressing gasket edges.*

special notes inside each box relating to specific installation instructions. If your project takes a few months to complete, you may forget how certain assemblies were dismantled, so notes explaining how they go back together may be heaven sent. For extended storage, fold top flaps over boxes or cover openings with towels or other material. This helps to prevent dust and moisture accumulation on top of parts.

If your storage facilities are small and will only accommodate boxes of parts, you will have to find a way to protect the open engine compartment during restoration. This can be done with tarps, large sheets of heavy plastic, or a makeshift enclosure arrangement. Should the firewall, inner fenders, and other painted body parts be stripped to bare metal in preparation for new paint, you might have to work in sections. This way, after each area is individually stripped, it can be properly prepared and then coated with a rust-inhibiting epoxy primer right away to minimize potential rust problems.

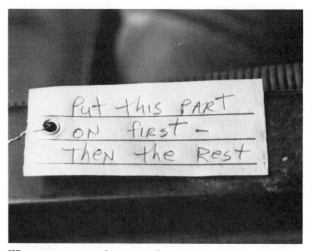

When parts are taken out of an engine compartment, a systematic means of parts storage must be maintained. If not, future retrieval and installation could be difficult. Parts tags, like this one, work great, especially when special notes or instructions are included for future reference. Tag courtesy The Eastwood Company

Along with adequate lighting, power and water sources, and vehicle and parts storage, your work site should allow ample room to maneuver. Park your car or truck only partway into a garage to make use of open space between the front of the vehicle and the back garage wall. If your garage is poorly lit, back your car into it and take advantage of any sunlight that shines in through the open door. Avoid the urge to quickly throw open the hood, though, as it could crash into a low garage or carport ceiling. With a low workplace ceiling, the hood might have to remain partly closed. Since you may not be used to working around your car's engine compartment with its hood hanging down slightly, be alert at all times to avoid bumping your head.

Although your choice of work sites may be limited, you should be able to make suitable accommodations if you plan ahead. Take some time to arrange a workplace before starting detail work. Set up a portable workbench by laying a sheet of plywood on top of a couple of sawhorses, or arrange tools, equipment, and materials on top of a picnic table. Reaching all areas around an engine compartment may be difficult, so have a step stool or heavy-duty five-gallon bucket handy. Having all of the necessary tools and supplies close at hand will help your project to progress smoothly and allow you time to concentrate on the work instead of having to constantly break away to hunt for a tool, brush, paint gun, air hose, rag, or whatever.

Cleaning Solvents and Soaps

Autobody paint and supply stores and auto parts houses generally carry wide assortments of engine degreasers, general cleaners, and car wash soaps. In addition, auto-related periodicals like *Hemmings*

The entire body of Dan Mycon's 1948 Chevy received extensive body and paint work. Note that the firewall has been customized to a flattened-out appearance. Anytime paint is removed from body panels, bare metal has to be chemically treated to remove formations of rust, and then prepped with an epoxy primer material to prevent rust from getting started again. Mycon used PPG's DP 40 Epoxy Primer on the engine, frame, suspension, and body before painting.

Pouring engine solvent from a gallon can is neither easy nor economical for engine cleaning. Use a plastic spray bottle or something like this handy refillable aerosol sprayer from The Eastwood Company. An air compressor can fill it with from 90 to 100 pounds of pressure and it will spray degreaser products, rust treatment chemicals, or any liquid with a viscosity up to 10SAE (Society of Automotive Engineers measurement). Sprayer courtesy The Eastwood Company

Motor News carry advertisements from mail-order companies that offer all sorts of automotive cleaning products. There are even special products designed just for stainless steel braided lines.

There doesn't appear to be any miracle degreasing or cleaning products available that can simply be sprayed on and rinsed off, producing perfect results after just one application. But there are a number of products that work well and have satisfied many customers for a long time. If you haven't quite found the right combination of engine cleaners for your needs, talk with other enthusiasts about the success they have had using various brand-name items or other materials.

While some detailers may deem kerosene or generic engine solvent as good degreaser, others may have had better results using degreasers off the shelf of a local auto parts store. Try a number of products, if necessary, until you find the ones that work best for you. Be sure to read and follow product label instructions before spraying or pouring any cleaning agent on the engine or surrounding area of your car. Read and adhere to all special warnings and wear rubber gloves, eye protection, and any other safety equipment recommended.

Mild petroleum-based products like kerosene and cleaning solvents are generally considered safe to use on most engine parts for removing heavy grease concentrations. Most store-bought engine cleaners are made of petroleum distillates and are also advertised as safe and effective for normal applications although some plastic parts may be adversely affected, which is why many products carry warnings that recommend testing on a small area first.

Kerosene and cleaning solvents are generally available at local gasoline stations. Bring along an empty gallon can approved for flammable liquid use, as bulk material is stored in large drums and poured directly into smaller containers; gas stations do not normally package engine cleaners in carry-out containers. Cost is nominal, usually around $1 or $2 per gallon. If local gas stations do not carry such supplies, look under the headings of "Kerosene" or "Solvents" in the yellow pages of your telephone book.

Brand-name engine cleaner application instructions often recommend that users scrape off heavy grease accumulations before contents are sprayed, poured, or brushed on dirty surfaces. Should you

A number of brand name engine cleaners and degreasers, like Gunk, are available at auto parts stores. All seem to do a good job of loosening grease and grime deposits, especially when they are assisted by the agitation of a stiff parts brush or floppy paintbrush. Be sure to read label instructions and follow recommended safety procedures before using degreasers.

Superchargers and turbochargers get hot during engine operation. Washing them immediately after the vehicle has been driven is a mistake, as cool water poured directly on their hot outer surfaces could quickly cause a fracture. Instead of relying on water flow for cleaning, allow blowers of all types to cool off to ambient, or room, temperature and then clean them by hand with soft cloths and mild cleaning agents, like Simple Green.

A clean cloth was gently positioned in the throat of this carburetor, then the entire opening was covered with a thick, plastic freezer bag held in place with masking tape. The bright spot to the left of the carburetor is aluminum foil covering the distributor. A degreaser product was sprayed onto the base of the carburetor and is now being agitated by a paintbrush to loosen accumulations of grease and grime.

If you have the facilities for major engine cleaning but lack the resources of a power washer, this Grime Blaster from The Eastwood Company may help. Used in combination with air pressure from a compressor, this unit can even siphon solvent so it can be applied at the same time as air pressure and water. Photo courtesy The Eastwood Company

need to do this, consider using paper towels to wipe off big globs and then a wooden or plastic paint stir stick to scrape off more stubborn build-up. This should prevent unnecessary scratches on painted or polished engine or compartment surfaces. A metal putty knife might work well but could easily scratch paint, cut wires, nick hoses, or cause other damage.

Most degreaser application directions also mention that engines are cleaned more easily when they are warm but, and the words are highlighted, *not hot.* Although this is true, owners of supercharged and especially turbocharged engines must be careful. Blowers get very hot under normal operating conditions and the sudden flow of cold water can quickly cause fractures and other serious damage. Those engines equipped with blowers of any kind must be washed or cleaned *only* after they have cooled off to ambient or room temperature.

Once a degreaser is applied, through whatever means, use a stiff parts brush or old paintbrush to work material into accumulations of grease and grime. Following label instructions, allow degreaser about ten or fifteen minutes to soak in. Then, rinse it

The thin layer of dirt that covered this Volkswagen Jetta engine compartment was easily and quickly removed with garden hose water pressure and help from Simple Green, a paintbrush, a detail brush, and a wash mitt. Note the plastic wrap that covers the distributor in the lower right side of the photo. Simple Green has been used by many auto enthusiasts with successful results.

After a road trip, Jerry McKee will fill a bucket with car wash soap and clear water and then use any assortment of items to clean the engine compartment on his Jaguar E-Type. Featured here is an array of cleaning brushes, Simple Green cleaner, masking tape and plastic wrap to cover electrical items, a soft cotton wash mitt, and assorted polish, wax, and dressing products by Meguiar's.

off with a heavy stream of water. Repeat applications until satisfied with results. Be sure to frequently rinse fenders and all painted surfaces when using an engine degreaser. Allowed to dry on exterior body surfaces, these products could remove wax protection and possibly streak body paint. Once an engine compartment is degreased, start the motor and let it run for about fifteen minutes or until dry.

Car wash soaps are available at autobody paint and supply stores, auto parts houses, and even a number of discount, hardware, and convenience stores. Of course, all claim to be the best, and most do a good job of cleaning automobile bodies. You will have to experiment with a number of different products to find the one that works best on your car. Art Wentworth prefers to use a mild liquid dish soap on his cars, believing that if it is mild enough for dishes and hands, it must be gentle on car bodies, too. He always uses a liquid to avoid any scratch possi-

bilities. Although most granulated car wash products do a fine job for a lot of satisfied customers, he feels that just one undissolved soap granule stuck in a wash mitt would pose too much of a scratch hazard.

Mild car wash soap thoroughly mixed in a bucket of clear water makes a great cleaning solution for washing engine compartments that are just slightly soiled. A soft, cotton wash mitt full of a sudsy soap solution goes a long way toward cleaning painted inner fender aprons, firewall surfaces, radiator supports, valve covers, air cleaner assemblies, and the like. Complement wash mitt use with brisk agitation of a soft, floppy paintbrush and you will be amazed at how well your cleaning efforts will be rewarded.

For more stubborn cleaning chores, you will have to use a stronger cleaner. Many enthusiasts have had great results using a product called Simple Green. It is a concentrated liquid cleaner advertised as "non-toxic, biodegradable, non-abrasive and contains no harmful bleach or ammonia." Wentworth, Mycon, and McKee are very happy with the way this product works for removing dirt and grime from all parts of their special cars and motorcycles.

After a long road trip in his 1972 Jaguar XKE Roadster, McKee will mix a few ounces of Simple Green with water in a bucket, dip in a soft, cotton wash mitt, and bring up a sudsy lather to clean the fenderwells and other areas around his car's V-12 engine. Stubborn areas are saturated with Simple Green directly from a spray bottle and then washed with a wet mitt. Although more than one application may be needed for extra-stubborn spots, overall performance of this cleaner has been very good. For the best value, purchase the product in gallon containers and then simply fill handy squirt bottles as they empty.

Many other general cleaners are used by auto enthusiasts and concours winners. Some won't use anything but Dawn dishwashing liquid, others like all-purpose cleaners found in supermarkets, and many rely on other name-brand cleaners specifically made for automotive use. Companies that used to have only one or two cosmetic car care products on the market now carry a number of items from multipurpose dressing and cleaners to polishes and upholstery shampoo. Once again, you'll have to talk over these choices with other car enthusiasts and also try a number of them until you discover which work best for you.

Whichever product you try, always read the label carefully and follow application instructions. These days, cleaning products seem to be much more high-tech than the old, general-purpose standbys of years past, with a number of them specialized for one or two specific applications. This should be a concern for owners of automobiles that sport highly polished alloy parts, billet add-ons, neon-colored wires, and other custom attributes.

Tools and Brushes

Automotive tool and equipment outlets, like The Eastwood Company and Griot's Garage, generally carry large selections of various detailing items.

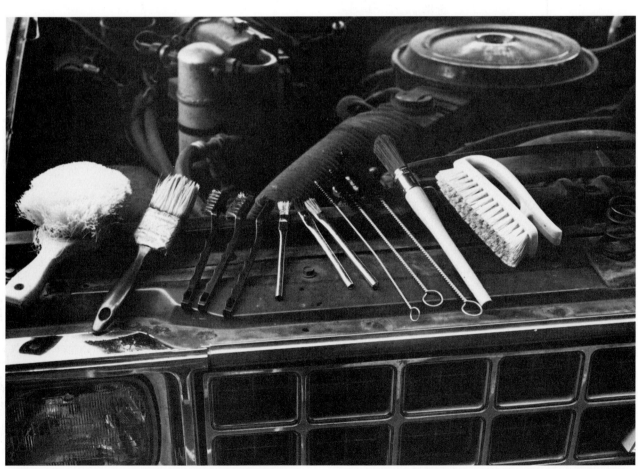

This typical assortment of engine detailing brushes comes in handy for certain tasks but is not required for every engine detail job. Starting from the left, a common plastic-bristled brush, a soft paintbrush, a set of three detailing brushes (each with different bristle strengths), three small parts brushes, three cylindrical engine cleaning brushes, a stout parts brush, and finally another type of common plastic-bristled brush. Nine center brushes courtesy The Eastwood Company

Their catalogs commonly offer things like refillable hand-held spray pots, water pressure wands, wash mitts, brushes in a variety of shapes, sizes, and bristle textures, Turkish towels, detail paint guns, spark plug plugs, and so on. Although you do not need to have all available detail tools at your disposal, a small assortment will help almost any engine compartment detail task progress faster and more efficiently.

In-depth cleaning and detailing most often requires use of some mechanic's hand tools to remove parts like air cleaners, hose brackets, windshield washer fluid containers, radiator overflow containers, batteries, and other items attached to structures around the engine compartment. You should have a quality set of wrenches and sockets in both US and metric sizes for American cars and trucks; many models seem to require a combination of both, even 1980 Chevrolet Suburbans. You'll need screwdrivers with crisp blades, both slotted and Phillips. And, if you plan to remove some of the grille for access to areas behind it, you might need a set of Torx bits—

asterisk-shaped screwheads that are becoming more and more popular.

A number of other specialty tools are available for engine compartment restoration work and custom modification. Flaring tools and tube benders are important items when rerouting fuel, brake, and automatic transmission cooler lines. Restorers working on neglected antiques may want to use thread chasers to rejuvenate special screws and bolts that might be almost impossible to duplicate, as OEM replacements may be virtually nonexistent. A die grinder with assorted carbide burrs might be perfect for smoothing rough edges on an engine block scheduled for new paint and installation into a super custom street rod.

Engine compartment detailers and restorers can easily find the necessary tools and equipment in mail-order catalogs, or from hardware stores, auto parts houses, autobody paint and supply stores, or at auto swap meets. Prices vary from company to company, as does equipment quality. Shop around and compare and don't always opt for the least-expensive models. Cheap detail paint guns, for example, may work great the first time but might fail to spray a tight, uniform pattern the next. Dan Mycon recommends spending the extra money for a quality Binks, Sharpe, or DeVilbiss paint gun because they will last longer and their rebuild kits are easy to find and install.

A variety of mechanic's hand tools will be needed to remove engine compartment parts such as the air cleaner, battery, radiator overflow container, and other items. Foreign-made vehicles will require metric tools, and domestic automobiles may call for both US and metric sizes. Be sure to have sturdy slotted and Phillips screwdrivers available, too.

Special sockets, like these from The Eastwood Company, come in handy when trying to remove an oil pressure unit or vacuum switch while painting the engine. In lieu of removing such items, try patiently masking with quality automotive masking tape to protect them against unsightly overspray blemishes.

This set of Torx bits fit into an adapter designed for use with a ⅜in ratchet. Torx fittings and fasteners are becoming more and more popular, especially for front end auto assemblies such as grille pieces, lights, and so on. Do not try to remove Torx head screws with a Phillips screwdriver or anything other than a Torx bit, however, as you will surely strip out the screws and be stuck with a more involved repair job. Torx bits courtesy The Eastwood Company

Meticulous detailers always have a good assortment of cleaning brushes on hand for engine compartment work. A large, floppy paintbrush is great for cleaning around assemblies mounted to inner fender aprons and firewalls. A small, one-inch-wide paintbrush is better for tight work on intake manifolds and carburetors. Stiff parts brushes make quick work of dislodging baked-on grease build-up problems, and small wire brushes may be needed for tough pockets of dirt and grime on front suspension pieces.

A variety of detailing brushes are available through auto tool and equipment suppliers such as The Eastwood Company. As much as these items can help make cleaning work go faster, those on a tight budget may have to settle for something less professional, like old toothbrushes. They work fine for many of the detail chores, but do not generally hold up as well as those specifically designed for engine detailing. Therefore, you may need to have two or three on hand. Use a toothbrush to clean around the edges of brake fluid reservoirs, valve covers, carburetors, intake manifolds, and so on. There are no clearcut rules for toothbrush use, so feel free to be inventive and use one wherever you deem necessary.

Eastwood's ISO Flaring Tool is not a standard engine detailing item. However, those who are customizing a project car or truck might find this sort of tool valuable when it comes time to install custom tubing. Tool courtesy The Eastwood Company

For bigger cleaning jobs, you may want to use any one of assorted plastic-bristled brushes commonly found in the household supplies section of most supermarkets. These work great for removing dirt and road grime from plastic windshield washer containers, frame and suspension members, engine

A set of rethreading die units might be useful when trying to clean or repair damaged spindles or special bolts. Although they are not designed for cutting new threads, these rethreaders can straighten existing threads without damaging or undercutting them. They are an asset for engine rebuilders and restorers. Rethreaders courtesy The Eastwood Company

blocks and manifolds, rubber splash shields, hood hinges, and so forth. Used along with a bucket of sudsy wash soap and a squirt bottle of cleaner, plastic-bristled brushes will remove a lot of debris in a hurry. Use caution around polished alloy parts, though, as stiff bristles could scratch delicate surfaces.

Paint Products

The durability of newer catalyzed urethane paint systems to withstand high engine operating temperatures has enabled auto enthusiasts to paint engine blocks just about any color under the sun. Looking through magazines like *Hot Rod*, *Sport Truck*, and *Car Craft*, you'll notice that a number of custom car buffs are now painting entire engines the same color as the body. Mycon has done this to the engine in the 1948 Chevrolet he is putting together. Its block is black and its valve covers are Mycon Purple, identical to the lower part of the body.

This innovation has brought excitement to a wide range of auto enthusiasts. Now, those who like to customize their rides are not limited to a few high-temperature engine paints, but can be creative with vivid colors and lots of add-on equipment that also come in a wide assortment of tints and shades. The trick is to blend engine, wiring, firewall, inner fender aprons, radiator, and accessory colors to appear uniform and balanced, with hues complementing each other.

The Universal Thread Cutter, a compact tool that will clean and straighten damaged threads on bolts or studs, even in tight places on engines, frames, or suspension members. It works on all threads, fine or coarse, up to ¾in diameter. This tool would help restorers having trouble finding replacement bolts or threaded parts. Thread cutter courtesy The Eastwood Company

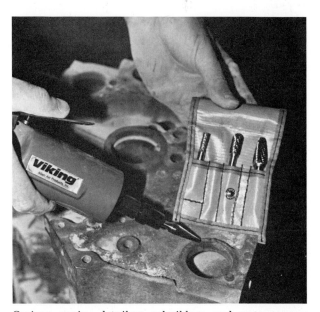

Serious engine detailers, rebuilders, and restorers use carbide burrs on a high-speed die grinder to grind weld beads, shape parts, and smooth rough parts in preparation for paint work. This set includes cylinder-shaped, pointed, tree-shaped, and round-nosed carbide burrs. Be safety minded by wearing eye protection whenever working with high-speed metal filing tools. Set courtesy The Eastwood Company

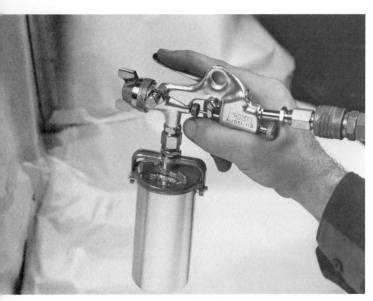

The Binks Touchup (detail) spray paint gun is an excellent tool for painting in tight spots such as engine compartments. Along with its handy size, the trigger mechanism allows for easy control by using the index finger. Gun cups are generally sold separately. The Eastwood Company

Along with high-temperature engine paints and durable urethanes, engine detailers must consider options for painting other engine compartment components. The Eastwood Company offers Stainless Steel Coating for exhaust manifolds and headers. It is long lasting and will hold up to temperatures as high as 1,200deg Fahrenheit. Other paint products such as Spray Gray, Detail Gray, and Aluma Blast are advertised as being identical to factory treatments for cast-steel parts, stamped and machined parts, and cast-aluminum parts, respectively.

Along with urethanes and high-temp paints, autobody paint and supply stores also sell other paint products designed for under-the-hood use. Color options are selected on an as-needed basis. Custom car enthusiasts may want anything but stock colors, while restorers of original machines must insist on factory-specified tints only. The acquisition of factory paint colors is critical to concours competitors and those who want to restore automobiles to factory originality. Once a color code is brought to an autobody paint supply specialist, mixing is seldom a problem. Most often, the trouble lies in locating factory-original color codes.

Original chassis black for many older cars, for example, is not gloss black. Many are flattened to specific percentages. An autobody paint and supply jobber can easily add a flattening agent to a gloss mixing black, but amounts added must correspond to factory specifications. Serious restorers spend lots of time researching shop manuals and other literature that pertains to their specific year, make, and model

This small parts brush is just the right size to clean inside the slots on these battery caps. It is also perfect for cleaning dirt and debris out of seams and grooves on other plastic parts like radiator overflow containers and windshield washer jugs. Brush courtesy The Eastwood Company

A small pressure washer works great for blowing off accumulations of grease and grime from engines, as well as for cleaning dirty wheels, fenderwells, and front suspension members. Units like this can be rented for a nominal fee. Be aware that the closer the nozzles are held to surfaces, the more pressure exerted and the easier it is to blow off decals, stickers, emblems, and the like.

until they locate the needed information. They will also check with experienced car club restorers and other experts in the field, like concours judges and professional auto restorers.

When it comes to serious engine compartment detailing or restoration, no item should be taken for granted. Eager novice concours competitors in a hurry to paint an engine part with something other than an original color or finish gloss will quickly learn the hard way at judging time: an experienced judge will recognize at once excessive gloss or an inappropriate color and note such on a judging score sheet. With concours events as keen and competitive as they are today, the loss of even tenths of a point for too much gloss or a color shaded too far off center can mean the difference between Best of Show and third place.

Special Equipment

The use of a pressure washer for cleaning engines is debatable. Water forced into electrical connections, alternators, air conditioning compressors, and other engine attributes is not something Detroit recommends every car owner do on a consistent basis. And for good reason. Sooner or later, that water forced into places it was not intended for will start a corrosion process that may involve lots of work to correct.

On the other hand, for an older daily driver with no intrinsic value other than normal commutes, it stands to reason that continued exposure to high engine operating temperatures would surely cause any water accumulations to quickly evaporate. You might not get much of an argument from the daily driver, but I doubt you could convince Concours

d'Elegance winner George Ridderbusch to use a pressure washer on his prized Porsche.

A choice between using a pressure washer on a daily driver or a concours winner obviously pits two extremes against each other. On one you would use high pressure, and on the other you wouldn't dare. But, what about a special 1960s vintage muscle car that you just bought from a retired farmer who drove it regularly, changed the oil and filter every 3,000 miles, maintained everything he was supposed to, but never cleaned nor had the engine detailed?

Art Wentworth has a pretty good answer for that question: "I love muscle cars just as much as I appreciate everything else with four wheels and would shudder at the thought of using a pressure washer on something I could easily clean by hand, even if it was going to be really labor intensive. But, I wouldn't have a problem using a pressure washer *one time* to get a special car cleaned up to the point where I could finish and then maintain it by hand."

That may be wise thinking for some, and maybe a mistake in judgment by others. Hundreds, if not thousands, of automobile engine compartments are steam cleaned or pressure washed every day by professional detailers and general auto enthusiasts. One could surmise that if using such a unit was so destructive, there would be widespread information passed along to every detail shop in the country to stop doing it or be prepared to pay for a lot of damaged or ruined engines.

Therefore, in the case of a greasy, filthy, grime-encrusted, and totally disgusting engine compart-

Moisture on the inside of a distributor cap will likely prevent the engine from starting. If it does start, chances are good it will run very rough. Regular blow dryers work great for evaporating moisture from distributor caps. Prolonged use could damage plastic caps, however, so refrain from holding dryers in one position for too long. Let your hand be a heat-sensing guide and sweep the dryer back and forth until all moisture has evaporated.

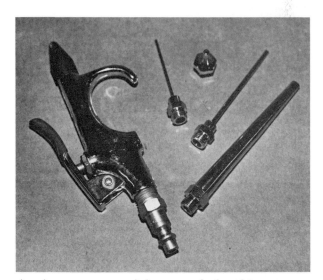

An air compressor unit is commonly found in many an auto enthusiast's garage and shop. While detailing an engine compartment, a Blow Gun set like this may be useful for dislodging debris in hard-to-reach pockets, cracks, and crevices. When using air pressure to remove dirt or other particles, consider wearing goggles to prevent contaminants from entering or injuring your eyes. Blow gun set courtesy The Eastwood Company

Jerry McKee is proud of his innovative approach to touching up bolt heads on his Jaguar E-Type Roadster's V-12 engine by using cotton swabs. He claims that they are inexpensive to use, don't have to be cleaned afterward, and don't drip paint off the ends. Other detailers have used artist's paintbrushes and the clean end of paper matchsticks.

ment, even on a special car, the use of a pressure washer one time may certainly be justified. With that, consider renting a small unit from a rental yard (when working on a special car) in lieu of taking the vehicle to a self-serve car wash. This way, you will not be under any automated timer constraints and can take all day, if need be, to carefully aim and operate the pressure washer wand exactly where it is needed.

Having a separate unit at your disposal allows you to meticulously pressure wash only those parts that need it and carefully avoid blowing off delicate paper decals, emblems, and stickers. At the same time, you can easily prevent high-pressure water concentrations from being forced into electrical connections, the distributor, carburetor, and other electrical and engine components. Small pressure washers are available at most rental yards and shouldn't cost much more than what you would have had to spend at a self-serve car wash.

Anytime you pour water over an engine compartment, you run the risk of getting water into the distributor cap area. As any car person knows, it doesn't take much more than a drop of water inside a distributor cap to render an automobile unstartable. In situations where water will be used, have a hair blow dryer handy. These gadgets work great for drying out small parts such as distributor caps. Do not hold the units too close, however, as excessive heat build-up can damage plastic caps. Simply wave a blow dryer over it for a few minutes until all signs of moisture have clearly evaporated.

Air compressors are fine machines that can be used during engine detailing to blow away dust, light dirt accumulations, water puddles, and a host of other things, including bug debris from inside radiator fin areas. The Eastwood Company offers a handy blow gun set for air compressors that could be quite useful during engine detailing. Along with a standard blow gun assembly, a series of four pinpoint nozzle tips can be attached to help guide thin but powerful air-pressure streams into extra-tight nooks, crannies, and crevices.

A quality air compressor is also a mandatory piece of equipment when planning to use a spray paint gun or any number of pneumatic power tools like a die grinder, high-speed rotary sander, dual-action sander, or air ratchet. If you don't own or have access to a quality air compressor, you can rent one from a rental yard. Fees vary and are calculated by size and cfm (cubic feet per minute) capacity. Be sure to check the cfm rating on pneumatic tools to make sure the compressor can properly supply them. This is especially important for paint guns. Inadequate air supplies will cause problems in obtaining perfect spray patterns.

Innovative engine compartment detailers have used a number of different tools, materials, and equipment to help them successfully and meticulously complete jobs. Wet-and-dry vacuum cleaners have been enlisted to remove stagnant wash water from intake manifold areas and other spots where it had accumulated. Artist's paintbrushes are used to carefully paint lines on valve covers, air cleaners, and other engine parts. Cotton swabs come in handy as makeshift paintbrushes for touching up bolt heads and chips on other items, as have paper matchsticks. Wooden toothpicks are favorites for removing fine lines of grime build-up along extra-thin grooves and screw slots.

Plenty of absorbent rags are always needed and old cotton T-shirts and flannel shirts make good ones, as do cloth diapers. Paper towels are handy for wiping away globs of grease because they can amass a good amount of debris and then be thrown away. Plastic food wrap and aluminum foil work well for covering distributors and carburetors during a wash, and automotive masking tape is essential while painting engine blocks and other parts. A small section of lightweight cardboard, like the bottom of a shoebox, works great as a paint block, and the nozzle and thin tube from a can of WD-40 lubricant works great for aiming and applying paint from a spray can.

To be sure, there are plenty of available detailing tools at your immediate disposal. All you have to do is plan your job and determine what it will take to complete each task. If you have never done anything like this before, you may be wise to watch a professional detailer perform his or her magic once or twice. Better yet, practice on an old sled first, like a friend's car, before tackling the engine compartment on your own special vehicle. Practice makes perfect, and after you have detailed a few engines by yourself, you'll have no trouble gathering and using all of the right tools, materials, and equipment to complete jobs efficiently and with professional results.

Chapter 3

Basic Degreasing

Every square inch of an engine compartment must be clean before detail work can continue past baseline. View engine detailing as a two-phase process; the first phase concentrates on overall and in-depth cleaning, and the second phase revolves around paint work, polishing, restoration, and the addition or replacement of decals/emblems, rejuvenated parts, or custom goodies. This chapter covers the first part of phase one: overall cleaning. Any novice engine detailer would be fooling him or herself if they thought minimal cleaning efforts could be covered up or masked with paint and the addition of brightly colored add-ons.

Besides the embarrassment caused by another

The addition of chrome valve covers or a custom air cleaner will not mask grease or dirt embedded in spark plug wires, on hoses, or along linkage assemblies. Therefore, time and effort given to cleaning tasks will pay off later when paint adheres smoothly to metal surfaces, and buffed-in dressing on rubber parts brings them to a perfect gloss.

auto enthusiast's ability to spot pockets of dirt, grease, or grime on a supposedly clean and detailed engine, those accumulations of crud cause other problems as well. Grease and oil attract dust, dirt, and other debris. Thus a haphazard engine cleaning might look halfway decent at first, but will quickly deteriorate as previously overlooked splotches of grease and oil multiply.

Cleaning engine compartments is messy work. Be prepared to get your hands dirty, as well as your arms, clothing, and shoes. Wear an old T-shirt or sweatshirt without any buttons; they present a scratch hazard while leaning over fenders. The same goes for pants; wear trousers without rivets or belts with big buckles. If need be, put on a long apron to eliminate any scratch hazards. Remove rings, watches, and other jewelry that could not only cause scratches but might also pose a safety hazard with regard to getting fingers stuck in tight places or grounding out electrical wiring.

Personal safety is an important issue for good reasons. Quite obviously, all fun would be taken out of an engine detailing project if you were to sustain an eye injury or other personal mishap. Be sure to follow all safety recommendations listed on labels of engine degreasers and general cleaners. Seriously consider wearing eye protection anytime you use potent cleaning or paint stripping chemicals, along with heavy-duty rubber gloves and a long-sleeve shirt. If a lot of sanding is planned for the removal of paint from body surfaces in an engine compartment, wear a dust mask or respirator. Think *safe*, be smart, and you will achieve all of your engine compartment detailing goals with complete satisfaction.

Removing Major Accumulations

Few car- or truck-loving enthusiasts would allow their personal rides to become so saturated with crud

It has been a long time since this firewall area has been cleaned. Removing such accumulations of grease and dirt build-up will be messy work. Plan on your hands getting dirty, as well as your clothes and shoes. Apply water streams at an angle away from your face to lessen the likelihood of debris ricocheting into your face. You should consider wearing goggles for such extra dirty work, especially when using potent degreasers.

Always remove rings, watches, and other jewelry when working on your car or truck. These kinds of items present scratch hazards, along with posing possible personal safety hazards should they get stuck between tight-fitting motor parts or electrical connections. Be aware that belt buckles and rivets on trousers are potential scratch hazards, too, as you lean over fenders while cleaning engine compartment areas.

that only major degreasing efforts could clean them. So, let's assume that you have just bought a used vehicle or are embarking upon an engine compartment detailing for a friend or family member.

Under normal operating conditions, engine compartment atmospheres are generally loaded with traces of petroleum-based products, coolant vapors, and a host of other airborne pollutants. Left unattended, the engine and engine compartment soon becomes covered with a light film of sticky deposits that attract dust, dirt, and road debris of all kinds. Before long, the entire compartment area becomes encrusted with thick blankets of grime that can only be removed by major degreasing methods.

The easiest and quickest method for removing large accumulations of grease and grime entails use of a steam cleaner or pressure washer. As discussed earlier, the location in which you tackle this job is important. Self-serve car wash facilities will require users to have plenty of quarters on hand to feed the machine. Operation time is limited through an automated timer, and you will have to work systematically to take optimum advantage of its short pressure

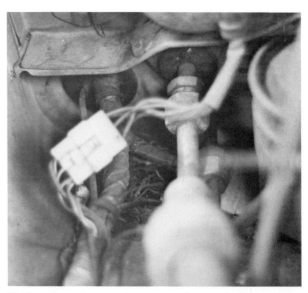

Neglected engine compartments frequently house accumulated debris of all kinds such as tree droppings and dead insects. Look in all corners for such material, especially at the base of the firewall and radiator supports.

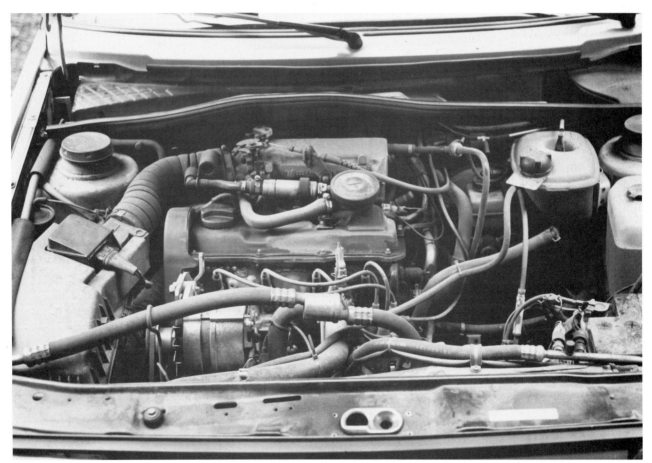

A somewhat heavy film of dirt and grease covers this 1988 Volkswagen Jetta engine compartment. Although the motor has never suffered a fluid leak of any kind, normal *engine operations create an oily substance that eventually covers all surfaces, which in turn attracts and holds dirt and road grime kicked up by the front tires.*

To prevent unnecessary short-circuit damage to electrical components in your engine compartment, plan to disconnect the battery before washing or cleaning begins. This is very important when planning to use high water pressure while conducting major degreasing chores. Note that the battery terminal, tray, and area below the steering linkage on this car need cleaning.

To prevent water or moisture from creeping into distributor cap areas, remove the cap and place a piece of plastic wrap over the rotor and entire inner body area. Let plastic hang out over distributor's outer edges. Then, replace the cap and secure clips or screws so they are held firmly in position. For additional protection, place towels or cloths on top of the entire unit.

washing cycles. Having a rented pressure washer at your disposal allows plenty of time to work at a leisurely and conscientious pace. If you can do this at home, turn up the hot-water heater temperature to 140 or 160deg Fahrenheit and use water directly from the heater. When the job has been completed, remember to turn the control back down to 120deg Fahrenheit.

If you drive your car to a self-serve car wash, you must avoid spraying water near its distributor cap, even when the cap is covered with plastic wrap, rags, or aluminum foil. If not, you run the risk of moisture penetrating into the cap space, a situation that might prevent the engine from starting and one that might leave you stranded with not only an engine that won't start, but a dead battery as well. Save distributor cleaning for later, when the car's at home and you have easy access to a hair blow dryer and electrical outlet, or an air compressor and plenty of time.

A thick, folded towel has been placed over this distributor after a piece of plastic wrap was laid over the internal surface area. The towel will help to prevent water from soaking into the distributor cap while major degreasing is carried out on the rest of the engine and engine compartment. The distributor will be cleaned later with damp cloths, small brushes, and a minimal amount of water.

Since water and electricity do not safely mix and because so many newer automobiles are equipped with lots of sophisticated electronic equipment, it is a good idea to disconnect batteries before spraying a large amount of water around the engine compartment. Likewise, you should refrain from drowning electronic components with high-pressure water to remove dirt or grease build-up. Diesel engines can be ruined by water in their fuel system. So, be very cautious when using water on cruddy diesels and direct water away from filters, injectors, and other fuel-related parts.

Pressure washing does a good job of knocking off coats of debris from hood undersides, engine blocks, and the like. However, to prevent unnecessary water penetration into electrical connections, alternators, computer boxes, and other such assemblies, use paper towels or rags to wipe off the big components and save the rest for in-depth cleaning later.

Although you may certainly add to it as you wish, the following is a general list of items you will probably need to satisfactorily degrease a neglected engine compartment:

Degreaser—a quart of liquid, or two to four cans of aerosol spray (two cans minimum for the engine compartment, one for the bottom of the engine plus an extra one).

All-purpose cleaner—a full 24oz squirt bottle or a smaller squirt bottle accompanied by a larger refill container.

Brushes—a floppy paintbrush, parts brush, plastic-bristled scrub brush or assorted smaller brushes.

Wash mitt—an old one saved for extra-dirty work that you can stick your hand into for prevention against cuts.

Rags—for cleaning and also to cover the distributor, carburetor, and electronic parts.

Paper towels—to remove large globs of grease and for possible use in covering distributor, carburetor, and so on.

Another way to protect the distributor during degreasing is with plastic wrap or aluminum foil. Here, a four-cylinder distributor cap has been covered with plastic wrap secured by masking tape. This system presents less bulk than a folded towel does, offering better access to areas around the distributor. Even with this protection, water should not be forcibly or heavily directed toward the distributor.

Loose-fitting or missing air cleaner wing nuts will allow water and degreasing products to run down support rods and directly into the carburetor throat. Eliminate this problem by removing the air cleaner housing lid and placing a large piece of plastic wrap over the carburetor opening and air filter area. Replace the lid so plastic fills spaces between it and the support rods.

Plastic wrap, aluminum foil, and masking tape—to cover the distributor, carburetor, and electronic assemblies.

Hand tools—to disconnect battery and remove air cleaner, windshield washer jug, and so forth.

Scraper—plastic or wooden paint stir stick or other nonscratching device.

Pressure washer—self-serve car wash, rental unit, or garden hose with a spray nozzle.

Bucket—to carry supplies and also to fill up with a soapy solution for dipping a wash mitt and brushes.

Towels—to cover fenders while leaning over them, protecting fragile components, and to dry large pockets of water after cleaning.

Pop open the distributor cap and lay a piece of plastic wrap over the rotor so it completely covers the inside of the distributor unit and hangs over the outer edges. Replace the cap securely. This will prevent water from seeping into distributor mechanisms.

After that, work a soft rag or paper towel between ignition wires on top of the distributor cap and cover it all with aluminum foil or plastic wrap. A rag or paper towel should absorb moisture that makes its way past the foil or plastic wrap. Use masking tape as necessary to hold it all in place.

An initial wash is started with the air cleaner in place; be sure top wing nuts are tightly secured to prevent water from entering the carburetor throat. Once the air cleaner is free of major accumulations, remove it to allow access to the carburetor and intake manifold. Gently stuff a clean, soft rag or small towel into the carburetor throat and then cover the entire opening with plastic wrap or aluminum foil. Some carburetors are designed in such a way that a plastic food container will snugly fit over them, provided a hole is made in the container to fit over the air cleaner support rod. Use masking tape, string, or large rubber bands to hold carburetor protection in place.

Some detailers prefer to wash the hood underside last to prevent lingering rinse water from drip-

The neck on this plastic part broke off inside a hose line, exposing both items to open air. Unless parts like this can be replaced with new ones before the start of a major engine compartment cleaning, use tape to seal off open ends to prevent water or degreaser from penetrating them. Larger parts may require that their open ends be covered with a piece of plastic wrap and then secured with masking tape or heavy-duty duct tape.

Large towels have been placed over the engine compartment to prevent water and dirt residue from dripping onto it while the hood's underside is washed. Once the hood has been cleaned, it will be towel dried to prevent any further dripping.

A clean sheet has been placed across a clean hood underside and secured with automotive masking tape before efforts got under way to clean the engine compartment. This prevented dirty water from splattering the hood while rinsing the motor and surrounding compartment. The windshield washer fluid container has been moved from its spot along the fender apron (right) to allow unobstructed cleaning in the area.

A build-up of grease and dirt like this is quickly loosened by applying a small amount of degreaser. Since the paint underneath is in excellent condition, this large accumulation will be easily and quickly wiped off with paper towels. Be very careful using paper towels on hood undersides, though; areas around holes and other openings may present sharp edges that will quickly cut fingers and hands.

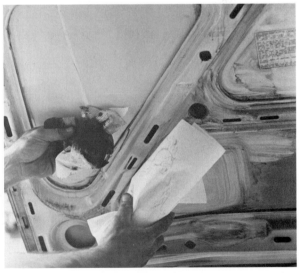

Paper towels are used to wipe away the bulk of accumulated dirt and greasy residue from the underside of this hood. Along with reducing the amount of dirty material that will ultimately be flushed into drains or sewers, this method prevents such residue from falling onto the engine or the towels draped over it. Paper towels will hold and absorb only limited amounts of crud, so plan to use several for this kind of work.

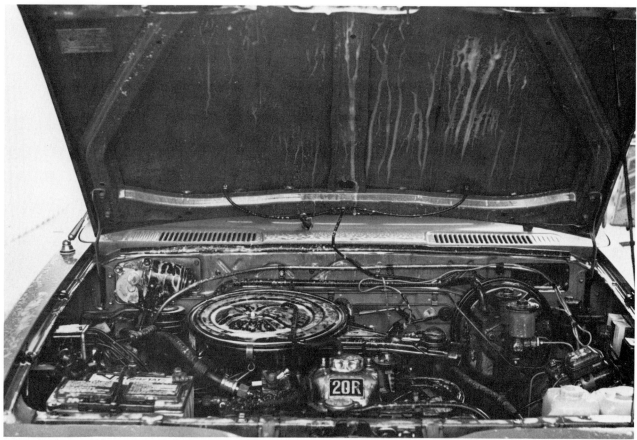

To remove large globs of crud heavily encrusted in engine compartments and on hoods of older, regular drivers, use a non-scratching scraper and paper towels. If you decide to skip that step, however, and rely instead on degreaser and high pressure water spray, remove as much of the big stuff as possible by soaking the engine compartment and hood with degreaser at one time. Give the degreaser time to soak in, then power wash the hood and compartment from top to bottom, and keep your head out from under the hood. This one-time power wash will give the degreaser a chance to penetrate much of the crud without rinse water diluting its strength when splashed onto surfaces from a previous operation.

Concentrated rinsing on top of an engine should include, but not be limited to all areas around the air cleaner housing, valve covers, accessible intake manifold areas, hoses, and clamps. Try to remove as much build-up as possible from bolt heads and inside grooves or indentations. The top part of this engine and nearby compartment spaces looks amazingly good after just two degreaser applications and pressure washer rinsings.

ping on their head while cleaning the rest of the engine compartment. This is fine, as long as you cover the clean engine compartment with a tarp before washing off accumulated debris from the hood. It might be better to clean the hood underside first, however, and then simply dry it off before proceeding. Tape towels or a large cloth to clean hood areas to prevent splashing grease and debris-laden water on them while working around other engine compartment areas.

Initially remove heavy grease and grime accumulations with paper towels or rags, enlisting use of a nonscratching scraper as you see fit. Then, either spray or brush on ample amounts of degreaser to all hood underside areas exhibiting greasy build-up. Allow degreaser to penetrate according to label directions, usually ten to fifteen minutes. With a pressure washer or garden hose, rinse the hood from top to bottom in a systematic pattern. Keep water away from insulation by holding nozzles close to work surfaces. Water-saturated insulation will hang down and drip for hours, tear, or simply fall off.

The hood has been washed and dried so drips of water do not fall on this detailer's head as he applies water from a self-serve pressure washer to dirty areas along the side of the engine. Poke your head inside the engine compartment to look for accumulations of dirt and grease while conducting major degreasing exercises. If you don't, the next phase of in-depth cleaning will require a great deal more work. Note that this detailer is wearing glasses for eye protection.

This 1988 Volkswagen Jetta is in excellent condition and therefore not a candidate for pressure washer cleaning. The bulk of grease and grime has been removed from the hood underside with spurts of degreaser and the use of paper towels. It was sprayed with Simple Green and hand washed using a soft, cotton wash mitt soaked with a cleaner solution from the nearby bucket. Rinse water was then applied through a garden hose at low pressure. Towels laid over the engine compartment will protect it from overspray. After an initial wash-mitt cleaning, more Simple Green was sprayed along the hood support to loosen build-up accumulated inside the tightly grooved space. A soft, floppy paintbrush works great for reaching into spaces like this to dislodge and break up encrusted dirt masses. Repeated efforts with a paintbrush and cleaner will help to gently remove grease, grime, or dirt accumulations in all sorts of tight spaces.

Once the hood has been thoroughly rinsed, move your attention toward the top of the engine. Rinse the air cleaner, valve covers, and other nearby components. Work your way down the front of the block being careful to avoid direct water spray onto the distributor, alternator, and other electrical assemblies. With the front of the engine rinsed, spray accessible areas to the right of the engine such as the brake and/or hydraulic clutch fluid reservoirs, windshield washer containers, inner fender apron panels, and so on. Do the same to the left side.

After you have rinsed as best you can from the front of the engine, step to one side and lean over the fender up to the hood edge. From that vantage point, you should be able to rinse the backside portions of parts at the front of the engine, like fan blades, shroud, grille supports, alternator or air conditioning compressor brackets, radiator overflow container,

and the like. Continue rinsing all areas that can be reached from that position, such as the side of the engine, exhaust manifold, firewall, steering linkage, frame, and so on. Rinse the other side of the engine compartment with equal vigor.

Cleaning Smaller Areas

The first degreaser application and following rinse should remove a great deal of encrusted grease, dirt, and grime. In some extreme cases, results may be

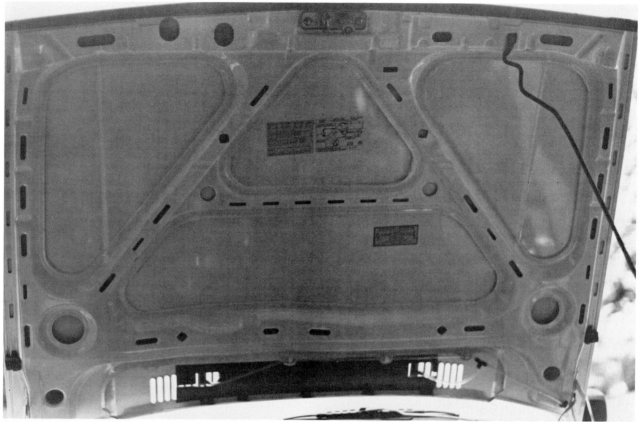

Simple Green, a soft cotton wash mitt, and plenty of paintbrush cleaning did an outstanding job of bringing this hood underside back to life. A soft towel was used to dry the surface in preparation for engine and compartment cleaning. A folded towel works well to absorb moisture that may be stuck inside grooves or other pockets. Again, be careful when working on hood under-

sides; lots of sharp metal edges protrude through holes and other openings. Remember what this hood underside looked like before? The job may have gone faster using a power washer and stiff-bristled brush, but the risks of blowing off stickers or creating swirl-like scratches on the paint were too great for a car in such good condition.

An old sheet has been draped across the clean hood underside and secured with automotive masking tape. This kind of tape will not damage paint that is in good condition (not flaking or peeling), nor will it leave behind a glue residue when removed. The sheet will protect the hood from splashes of water that will ricochet off surfaces when cleaning the engine compartment below.

This sticker used to be positioned an inch, or so, to the right of where it sits now, as indicated by the residual glue. A degreaser product was used to loosen the grime around the sticker and apparently loosened the glue as well. Water pressure from a simple garden hose was enough to move the sticker to its new location. The power of a pressure washer certainly would have blown it completely off. Therefore, pressure washer nozzles must be directed away from items like stickers, decals, and other semi-fragile accessories.

staggering. However, once steam has dissipated, you will probably notice that there is still a good deal of debris to contend with.

After completing the initial degreasing which encompassed the entire engine compartment and hood underside, you can begin concentrating on smaller areas, one at a time. First is the hood underside. Spray degreaser, as necessary, on stubborn accumulations of crud along just one half or quarter of the hood. Use a parts brush or paintbrush to work degreaser in and break loose crud. Rinse and repeat as necessary until the entire hood underside has been degreased, including the hinges.

Although degreasing in this manner normally proves quite successful, you should follow by using an all-purpose cleaner, wash mitt, and paintbrush to ensure the best results. Squirt a generous amount of cleaner, like Simple Green, over a section of the hood as you did with degreaser. Dip a wash mitt into your bucket of soapsuds and wash that section completely, using your fingertips inside the wash mitt to reach inside curved ridges and other tight protrusions. Many hood undersides exhibit sharp edges that will quickly cut bare fingers or hands, so be sure to keep your hand in the mitt while washing.

With plenty of suds still clinging to the hood underside, briskly agitate a soft, floppy paintbrush along edges, grooves, corners, inside framed openings, and anywhere you may have noticed accumulations of dirt or grime. Continue this process until the entire surface has been sufficiently cleaned. Don't forget hinges, sides, locking pins, and other small parts. Use clear water to rinse away soapsuds and dirt. Once satisfied with your cleaning efforts, use soft towels to dry the entire hood underside.

After the hood has been dried, use automotive masking tape to secure a lightweight tarp, old sheet, or other cover over the clean hood to protect it against splashing while you clean the rest of the engine compartment. *Note*: Use masking tape specifically designed for automotive use; generic tape often leaves behind glue residue after it is peeled off a painted surface.

From one side of the engine compartment, apply degreaser to accumulations of build-up as you see

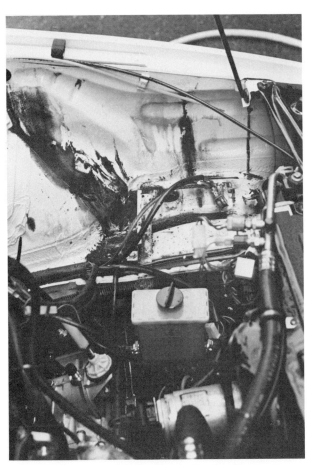

The accumulation of black crud on this fender apron was hidden behind a windshield washer reservoir. It is actually Cosmolene impregnated with grease and grime. Cosmolene is a waxy substance applied by car manufacturers to protect autobody parts from the effects of saltwater while being shipped across the ocean. This material was scraped off with a soft, plastic squeegee and small brush.

Look for grease and grime accumulations deep in your car or truck's engine compartment. Frame members frequently sport build-ups of dirt and debris, as do the tops of fuel pumps and suspension pieces. Use degreaser as needed, and accurately spray areas using a garden hose or pressure washer nozzle. Don't forget to use a parts brush or paintbrush to break loose pockets of crud.

them. Use a parts brush to scrub spots of stubborn grease and grime until they form a soupy consistency. A floppy paintbrush works best on carburetors. If a drop light is needed to improve visibility, be careful not to accidently spray it with degreaser. Hot light bulbs can explode when hit with cool liquids.

Rinse away greasy residue from the engine and all other compartment areas with plenty of clear water. Although holding a pressure-washer wand close to work surfaces will quickly remove stubborn build-up, be aware that even garden-hose pressure has enough power to blow off decals, stickers, and paint. Therefore, direct streams of water judiciously to avoid damage to stock decals on valve covers, batteries, identification labels on accessory parts, and so forth.

Inspecting the Engine Compartment

Upon completion of your second degreasing maneuver, make a thorough inspection of the entire

engine compartment. Look for pockets of grease on the front of the engine around the water pump, timing chain cover, and behind obstacles like the alternator, fan blades, and power steering and air conditioning units. On compartment sides, look carefully for accumulations of grease on suspension pieces, frame members, steering components, fuel and brake lines, along hoses and wiring harnesses, starter, fuel pump, and so on. Apply degreaser as needed and be sure to work it in with a parts brush or paintbrush for optimum results. Rinse with clear water.

Convinced that your engine compartment has been satisfactorily degreased, take a moment to look at the engine area from underneath. Since the ground will be wet, kneel on the garden hose to keep your pants dry. Apply degreaser as best you can to accessible spots under the engine compartment that exhibit signs of dirt or grease build-up. Rinse with water and repeat as necessary. An aerosol product or refillable pressure sprayer might work best for this job, since sprays generally have a better reach. Definitive degreasing and cleaning of bottom engine areas will require use of a pit, hydraulic lift, or heavy-duty car ramps.

Efforts so far should have removed the vast majority of grease from your engine and engine compartment. Next is a general wash with an all-purpose cleaner, wash mitt, and paintbrush, just like

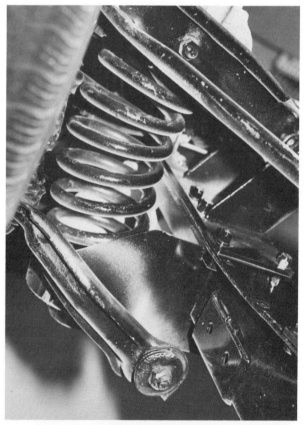

Up to this point, most of your work has been conducted while standing or leaning over your car. Don't forget to kneel down and check the condition of lower engine compartment assemblies. Springs, suspension supports, the oil pan, and many other components may suffer grease, grime, and dirt accumulations that can be easily removed with degreaser and water pressure. You should even consider cleaning those areas with an all-purpose cleaner, wash mitt, and paintbrush.

Getting dirt and crud out of tight areas along fender edges and other such body parts is made easy with the help of a soft detail brush, paintbrush, or toothbrush. Here, direct application of Simple Green and agitation from a soft detail brush makes quick work of removing crud from around bolt heads and along the top fender edge. A paintbrush should be kept close by so it can be used around the rubber hood bumper and elsewhere when needed.

you did to the hood. Wash only one small section at a time for best results. Spray a generous amount of cleaner on one half of the firewall, for example, and follow it with a sudsy wash mitt. Use a floppy paintbrush to work in soapsuds and dislodge dirt build-up along part edges, around grommets, under wires and hoses, in corners, behind heater units, and so on. Next, wash an inner fender panel using the same technique. Continue this operation until the entire engine compartment has been washed.

Only close visual inspection will reveal your general degreasing and initial cleaning progress. You must use a sharp eye to spot lingering spots of crud hidden behind various engine parts, hoses, wires, and other parts. The more intensely you hunt for and remove dirt and grease, the better your detail efforts will be rewarded.

Conscientious and concentrated use of potent degreasing products and pressure washers should leave your engine compartment very clean. Some novice detailers might stop at that and call the job complete; serious detailers would not. They realize that *wet* engine compartment assemblies frequently mask films of dirt and road grime that are not easily recognized until engine parts are *dry*.

Once satisfied that an engine compartment has been thoroughly degreased and initially freed from all obvious signs of crud build-up, remove protection from the distributor and carburetor, replace the air cleaner, and connect the battery. Start the engine and allow it to idle for fifteen to twenty minutes. If it won't start, check for moisture inside the distributor cap and dry it off with a hair blow dryer, air compressor, or clean, dry cloth.

When the engine starts, close the hood and let normal engine operating temperatures evaporate moisture from around the compartment. Keep a watchful eye on the engine temperature gauge to be sure the motor does not overheat. To speed this drying process, use a wet-dry vacuum cleaner to remove large pockets of water on the intake manifold and other compartment areas, or large towels that can absorb stagnant water.

Even though you should have continually rinsed the fender and cowling areas with clear water during degreasing work, this may be an ideal time to wash the front of your car from the top of the windshield forward. This step should guarantee that neither degreaser nor cleaner will dry on painted fender or cowling surfaces, and also afford you a chance to wash off any grit or debris collections that were inadvertently splashed on them.

Your cleaning progress can only be measured by close visual inspection. The unusual angles and shapes presented by this transaxle housing were cleaned with degreaser, Simple Green, a parts brush, and a paintbrush. Repeated cleaning efforts and visual inspections paid off; this housing is sparkling clean and ready for in-depth detailing.

The backside of a 1974 Jaguar XJ 12 L headlight, located on its hood structure which has been tilted forward. Notice the amount of dried dirt that has attached itself to the surface. Car wash soap, or a mild cleaner, combined with a soft wash mitt and paintbrush will quickly remove this eyesore to leave the area sparkling clean. Areas like this must not be overlooked. Always inspect spaces between grille fixtures and radiators for accumulations of bugs, road debris, tree leaves, and the like. Clean them as necessary.

Chapter 4

In-depth Cleaning

Since you probably will have relocated your vehicle from a self-serve car wash or other wash rack area to a garage or carport, remember to lay a sheet of plastic, piece of cardboard, or drip pan underneath

Although this Corvette engine looks clean, you can see light brown dirt stains featured along the base of the hood, the radiator supports, and inside the fan shroud. Cleaning oversights like this detract from an otherwise clean engine to make the compartment look less than desirable. These kinds of blemishes don't require a great deal of effort, just a little time and patience with a wash mitt, paintbrush, and some cleaner.

the engine compartment to catch globs of debris as they fall off during this next cleaning process.

Engine compartment areas generally will look great after a thorough degreasing and washing, especially while they are still wet. However, you may be disappointed when you open the hood after a motor has idled for some time and dried off most of the parts. They may not be as clean as you expected. Use caution when inspecting the area; the engine will be hot and so will its exhaust manifolds or headers, radiator, hoses, and other exposed items. If you need to lean on something while inspecting or cleaning, lay a thick, folded towel down and rest your arm on it.

General Cleaning

Once the engine compartment has dried you are likely to find light brown dirt stains on hoses and on other dark-colored parts. The engine will most likely look dull and faded with signs of dirt build-up along manifold gaskets, inside tight corners, around spark plugs, and anywhere else featuring grooves, ledges, or unusual shapes.

Begin by using a soft cloth and all-purpose cleaner on the hood underside. This way, dirt particles that may fall onto the engine compartment can be routinely cleaned later while working in that area. Fold a soft cleaning cloth into quarters to make it a manageable size, and also to provide you with a number of clean sides that can be unfolded as one becomes soiled. Spray cleaner directly onto the cloth and then wipe dirt away from surfaces. Follow up with swipes from a clean side of the cloth to remove moisture streaks.

Refrain from spraying cleaner directly onto work surfaces to keep overspray at a minimum and prevent accumulations of cleaner from becoming trapped inside recesses, as they will eventually drip or seep out to make a mess. Use a soft toothbrush or detail brush to dislodge dirt from sheet metal seams and other tight spaces. The folded edge of a cloth also works well to remove moisture and loosen dirt from those tight spaces. Be sure to clean *all* hood underside surface areas. This includes front, back, and side

Small pockets of grease or dirt build-up are commonly found along linkages and other small assemblies after a basic degreasing. The light film of greasy residue around mechanisms on this carburetor can be quickly removed with cleaner and agitation with a paintbrush. A cloth or wad of folded paper towels placed below this dirty spot will catch drips of cleaner and dirt residue before it runs all over the high-rise manifold.

If you are planning to conduct a thorough engine compartment detail or restoration on your special car or truck, complete with the removal of several engine parts, consider having someone videotape the parts removal process. This will give you an accurate reference when it comes time to reinstall the parts. In lieu of a video, take lots of photographs.

To save space inside vehicle interior compartments, many automobiles now carry jacking equipment inside the engine compartment. Some Subaru models even have a spare tire mounted under their hoods. In-depth cleaning requires the removal of jacking equipment, batteries, and other accessories to gain cleaning access to surfaces under them. Stow parts in a box (batteries on a piece of wood or rubber) for the time being, until engine compartment cleaning has been completed.

edges, as well as hinges and locking assemblies. Only when you are completely satisfied with cleaning results should you move on to other parts of the engine compartment.

If you plan to remove several parts from the engine or surrounding compartment area for super detail work or restoration, consider having someone videotape your dismantling process. As crazy as this may sound, you might be thankful to have a video recording of this sequence when you start putting everything back together. At the very least, especially for major dismantling jobs, take lots of photographs and notes for future reference.

Serious engine compartment cleaning requires removal of the air cleaner, windshield washer and radiator overflow containers, jacking equipment, and battery as a minimum. These items will be thoroughly cleaned and painted, if need be, out of the host vehicle. Their removal will also give you excellent cleaning access to areas they previously covered.

A soft, plastic-bristled brush is used to clean around the edges of a carburetor on McKee's E-Type Jaguar. This type of intricate work requires patience and the realization that even though it may be time consuming, it will mean the difference between a mediocre engine detail and a super one. The lettering on top of this Stromberg model sure looks good with all of the dirt and crud removed. Brush courtesy The Eastwood Company

A clean cloth was gently stuffed inside the throat of this carburetor to prevent debris from entering that space while work is conducted nearby. A small parts brush works great for dislodging grease and dirt from tiny corners and creases. Use a squirt bottle to apply degreaser or cleaner to parts like this. They work better than aerosols because the triggers allow for much better flow control and less overspray.

The engine compartment on this 1988 Jaguar is certainly packed full of parts. In-depth cleaning around this big motor is definitely hampered by a lack of space and accessibility. Situations like this call for innovative ideas.

Some parts may best be reached from underneath while the front end rests on heavy-duty jack stands. And removing a couple of key parts may provide more room for cleaning. Be creative.

There is no hard-and-fast rule that states you must begin cleaning at any one spot. You could start at the carburetor, work down the front of the engine, and then complete the sides, saving surrounding compartment areas for last. Or, you might prefer to begin at one corner of the engine compartment and work your way around to the opposite corner, saving the engine for last. The pattern itself is not nearly as important as maintaining a systematic approach, which guarantees you eventually cover every part of the engine and engine compartment. So, develop a plan that makes sense to you and stick with it.

Removing dirt and crud from tight spaces around the carburetors and those tiny parts attached to them requires patience and assistance from a paintbrush, toothbrush, or suitable detail brush. Cover carburetor throats with a clean cloth, plastic wrap, aluminum foil, or other material to prevent dirt particles or drops of cleaner from being flung into that area by brush bristles. Good results have been experienced when paper towels or rags are placed below the carburetor before carefully squirting concentrated cleaner on dirty spots and agitating with a paintbrush. The paper towels or rags catch dripping cleaner and dirt before they flow onto the intake

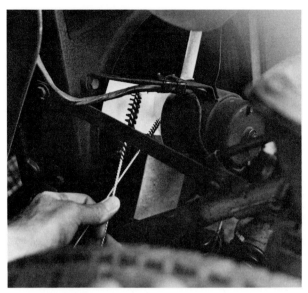

Engine detailers are always on the lookout for new ways of accomplishing difficult cleaning tasks. These brushes were originally designed for cleaning engine ports and passageways made accessible during complete engine rebuilding. Here, Wentworth uses them to clean around a fan motor on McKee's Jaguar. They work well, but one must be aware that their hard metal noses could quickly and easily scratch anything they come in contact with.

Heavy-duty jack stands are a must when crawling under an automobile for detailing, repair, or service work. Regular bumper jacks, scissors jacks, hydraulic jacks, or other types are not acceptable for safety reasons. Always rest vehicle bodies on sturdy jack stands before going under them for any reason.

The front part of an engine is generally filled with accessory parts, belts, hoses, and so on. These items make access to the block and water pump difficult. If you are determined to clean the front of your engine properly, consider removing parts such as the alternator, air conditioning compressor, belts, fan shroud, and fan. They can then be meticulously cleaned and detailed out of the compartment, where they are easier to manipulate.

mainfold. Use clean paper towels or absorbent cloths to wipe away residue.

Early on, you'll need a stout toothbrush or equivalent detail brush when cleaning the intake manifold. These handy tools do a great job of dislodging dirt build-up in corners and creases. Be sure to immediately wipe off residue as it collects in pockets featured along the sides of the intake manifold where it meets head and valve cover edges. Should you come across a patch of extra-stubborn grease or crud, use a little degreaser in lieu of all-purpose cleaner. Briskly agitate with a brush for best results.

Reaching engine side areas is not always easy, especially on newer cars which seem to have engine compartments crammed full of parts. Unless you know exactly what you are doing and have experience removing and installing such items as smog control devices, vacuum lines, computer sensors, and other high-tech components, you may have to settle for cleaning just those spaces that can be reached by hand, with brushes, or a combination of aerosol cleaners and a garden hose.

If you are intent upon cleaning the sides of your engine, consider raising the front end and resting it on sturdy jack stands before attempting to clean from below. *Never* rely solely on hydraulic or mechanical jacks to hold a car up. *Always* use heavy-duty jack stands and *always* block those tires remaining in contact with the ground.

Cleaning the front part of an engine requires a similar degree of patience as needed for detailing a carburetor. You are faced with a number of nooks and crannies, brackets and supports, with access usually obstructed by an alternator, fan shroud, and belts. If you have the time and ambition, remove the shroud, fan, and belts to gain better access. This will also give you a chance to meticulously clean both the shroud and fan while they are out of the engine compart-

As you move about the engine compartment on your car or truck, do not forget to clean cowling areas. Certain automobiles feature cowling spaces, like this, that are susceptible to airborne debris such as tree droppings, dirt, and mold. If needed, remove screens to clean spaces under them and to clear any openings that may be plugged with insects, leaves, or other debris.

Polish and wax residue has been building up in this corner and around nearby bolt heads for some time. Use a cutoff paintbrush to loosen the material. Then spray a little cleaner into the area and agitate it with a floppy paintbrush. Gently rinse with clear water or wipe away the residue with a soft cloth.

The owner of this 1958 Chevrolet has made a habit of keeping the area between the grille and radiator clean. These spaces tend to accumulate dust and road grime and need to be cleaned on a regular basis—like every time the car is washed. When washing or drying such surfaces, protect your hands against cuts by keeping them inside wash mitts or by wrapping them inside a towel or cloth.

Engine compartments are generally filled with assorted hoses and wires. Because the bottom portions of these parts are usually hidden, they are frequently overlooked during cleaning. Loosen hose holders or wire brackets to gain access to their undersides. At the same time, clean motor, firewall, or fender apron surfaces that are exposed once hoses or wires are moved.

Rubber and soft plastic or vinyl parts, like battery terminal covers, collect greasy films and oily deposits quite easily. Mild scrubbing with a soft detail brush or toothbrush works great for removing build-up from impregnated surfaces, and along cornered impressions and indented lettering. These items may be easier to work with after they have been removed from the battery or other parts.

ment. Make a drawing complete with notes on how belts are routed to be sure they are put back on correctly.

If you go to the trouble to loosen an alternator and/or air conditioning compressor to remove belts, consider taking them off completely. Their removal allows you to clean and detail them out of the car and also gives you better cleaning access to those engine areas under them.

The firewall, inner fender structures, and areas on both sides of the radiator need the same kind of meticulous cleaning. All-purpose cleaner and a soft cloth should work well on wide-open areas. Use a soft toothbrush or detail brush to dislodge concentrations of dirt in corners, seams, and creases. Be sure to move hoses and wires so you can clean the surfaces underneath them. Loosen hose and wire hold-down brackets as necessary.

Since many sharp sheet-metal edges are present along inner fender aprons and radiator-grille support

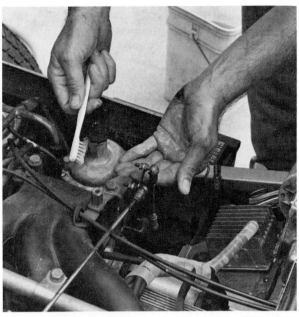

To keep cleaning residue at a minimum, Art Wentworth works up a paste of powdered cleanser and water in his hand and dips the toothbrush in it before scrubbing a carburetor piece. Powdered cleanser provides more cleaning strength to remove stubborn build-up that cannot be loosened with a liquid cleaner. This method works well in areas already cleaned but still showing signs of lingering dirt accumulations.

A soft, floppy paintbrush is one of the most versatile tools an engine detailer can own. It does an excellent job of cleaning rough-textured flat surfaces, such as wrinkle paint, as well as in corners, along grooves, and around a host of other tight spaces. Be sure to wrap thick duct tape around the metal band of your paintbrush to prevent accidental paint chipping.

The best way to clean spark plug wires is by hand using a soft cloth dampened with cleaner. One by one, remove wires, clean them, and then put them back on the engine. This way, there will be no confusion about maintaining their correct firing order. As one side of the cloth becomes soiled, unfold it to a clean section. Spray cleaner on the cloth, not directly on wires. This will keep overspray to a minimum.

This radiator cap was sprayed by a novice detailer. Note the black overspray on the front cross-member. Most paint overspray can be removed with lacquer thinner on a rag. Be careful, though; lacquer thinner is a very strong material. If you are not sure whether an underlying surface can withstand the effects of lacquer thinner, opt instead for repeated applications of paint thinner or a wax and grease remover. Note: Lacquer thinner will damage enamel paint.

The stock battery from John Pfanstiehl's 1959 Cadillac Eldorado. Even though the car only has 2,232 miles on it—that's right, less than 2,300 miles—it needed some refurbishing. Concours competitors and those restoring original automobiles must realize the importance of maintaining stock parts. Cleaned up, this battery will look great. A 50:50 mixture of baking soda and water should clean off the acidic residue once caps are securely put back in place.

The baking soda solution did a good job on top, and cleaner worked well on the sides of Pfanstiehl's battery. A soft brush and cleaner removed dirt and build-up from the rough-textured battery case. All that is left to do is brighten the battery terminals and this battery will be ready to show, along with its unbelievably low mileage car.

This engine compartment has more than 50,000 miles on it and yet it still looks good with just an in-depth cleaning. The amount of time you spend cleaning your car or truck's engine and engine compartment will have a direct impact on how well it sparkles when you're done. It is common for an enthusiast to spend days cleaning every square inch of a favorite vehicle's engine compartment, and the results, of course, are fantastic.

Remove the battery and windshield washer reservoir for definitive cleaning underneath them. Remember how bad that area looked when it was caked with grease-impregnated Cosmolene? It only took a few extra minutes to take those parts off, and the efforts involved in cleaning around them were minimal. This kind of in-depth cleaning helps to make the difference between a good engine compartment detail and a great one.

With all of the grease and grime removed, this little Volkswagen engine looks pretty good. Scrubbing with a parts brush, detail brush, and toothbrush helped to remove build-up from the head and the raised letters on it. A paintbrush worked effectively on the alternator, hoses, wires, and connectors. Plastic over the distributor did its job, too; the engine started on the first try.

units, drape a cleaning cloth over your hand for optimum protection from cuts and scratches. You might even consider using a damp wash mitt in lieu of a cloth. Follow cleaning with a dry towel to wipe away moisture streaks and expose spots that may have been missed. Remember, wet parts generally look clean and shiny but once dry, you can quickly see their actual condition.

A common oversight made by novice engine detailers is failure to clean the bottom sides of hoses, wires, and tubing lines. It is easy to do, since those sides are generally hidden from view. When you cannot see the underside of a part, rely on the amount of dirt that comes off on your cleaning cloth to determine when they are clean. In other words, they should be clean when you can wipe them off and no dirt residue is noticed on the cloth. For tight spots, feed an end of the cloth under the hose or tube and then buff it, as if you were shining a shoe.

Wires are semi-delicate items in that you cannot pull and tug on them too much while cleaning. Many detailers have had good luck placing a cloth over one hand, wetting it with cleaner, and then grasping the wire with the cloth. By securing the wire's end with your free hand, you can then slide the cleaning cloth along the wire's length for cleaning. This maneuver may have to be repeated a few times to achieve satisfactory results.

Hose clamps, tubing line connectors, and electrical connections must all be cleaned, just like everything else. Try holding a cloth under these items with one hand while scrubbing them with a toothbrush or plastic detail brush. The cloth prevents debris from falling onto other parts. Small electrical connections can be cleaned the same way. Squirt cleaner directly

A long afternoon of leisurely paced cleaning work with a garden hose, a little degreaser, some paper towels and soft cloths, Simple Green, a wash mitt, and assorted brushes helped to make this 1988 Jetta engine compartment look like new. Another afternoon spent polishing, waxing, and dressing certain items should make it sparkle like a show car.

It shouldn't be hard to remember what this 1976 Toyota engine compartment used to look like. Because of its filthy condition, pressure washer power was used to remove several years of caked-on grease and grime. Now, after four pressure washer sequences and a few hours of in-depth cleaning, this compartment looks as if it had belonged to a conscientious owner. The truck's resale value most likely has increased by much more than the cost incurred to clean the engine and engine compartment.

61

The following series of photos was taken of Jerry McKee's 1972 Jaguar XKE V-12 Roadster as he and Art Wentworth prepared it for a local Concours d'Elegance competition. Because the V-12 in McKee's Jaguar is very sensitive to moisture, water must be kept out of the valley as much as possible. Before using water to wash other engine compartment areas, towels were carefully folded and laid down over the components located in the valley. Next, a couple of newspaper sections were placed over the towels to absorb as much water as possible. Finally, a wide sheet of plastic wrap was positioned over the newspaper and then held in place with automotive masking tape. This should keep the valley dry while washing tasks are completed around it.

This dirty fan blade is located in an area difficult to reach. About the easiest way to clean it is with a spray bottle of cleaner and a floppy paintbrush. Likewise, the wrinkle paint finish on many parts will require use of a paintbrush or soft detail brush for maximum cleaning results.

To prepare an automobile for concours or a car show, engine compartment detailers must systematically hunt for and then remove all traces of grease, oil, grime, and dirt. Here, an oily film has been located on a front shock absorber, an item clearly within view of anyone looking into the engine compartment of this E-Type Roadster.

onto parts or brush bristles, whichever is most convenient and will cause the least amount of overspray.

Ignition wires should be cleaned individually. Use a mild cleaner and a cloth; do *not* dunk them in water. Take one wire off, clean, and then replace it before removing another. This guarantees that their correct firing order and position will be maintained. Use a soft toothbrush to remove build-up from ridges on spark plug and distributor cap ends. Ignition wire brackets and holders can be removed for cleaning in your bucket of soapy water with help from a plastic brush. Use all-purpose cleaner directly, as needed. And, don't forget to clean areas along valve covers exposed by the removal of those brackets or holders.

Distributor caps may be cleaned even with ignition wires snapped in place. Use a thin, long-bristled paintbrush or large artist's lettering brush and a soft cloth. Dampen paintbrush bristles with cleaner and swish it all around the distributor cap area. This should loosen dirt and turn it into a soupy consistency. Then, place the brush inside a piece of folded or wadded cloth and use the handle to wipe away dirt and cleaner residue. You can also use your fingers inside a cloth to achieve the same effect.

Should the engine compartment exhibit signs of paint overspray on wires, hoses, and so on from previous detail work, dab a bit of paint thinner or wax and grease remover on a cloth and gently rub to remove overspray. If this does not work, use lacquer thinner. Use caution, however, since paint and lacquer thinners are highly flammable. If you are working in a garage or shop, be sure there are no water heater or furnace pilot lights close by, and do not smoke while working with these substances.

Paint overspray on paper decals or stickers is almost impossible to remove without erasing printed material. Try to *carefully* remove overspray with a dab of paint thinner on a cotton swab or the corner of a folded cloth. Sometimes, a *very* light touch with #0000 steel wool will scrape paint overspray off without damaging printed lettering. If paper stickers or decals are damaged, you will have to replace them with new ones. Plastic decals and emblems, on the other hand, may fare a little better but are also susceptible to damage from thinner.

Lower engine compartment areas can be cleaned by reaching in from above. This is not always

Simple Green and paper towels were used to remove most of the oily film from a front shock absorber. Along with doing a good job of removing such residue without damaging underlying finishes, paper towels save cleaning cloths for more intricate work.

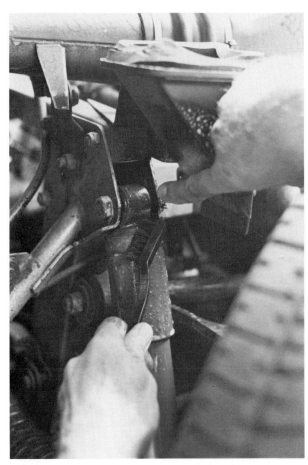

Getting road grime out from around a shock absorber mount requires a bit of scrubbing with a soft detail brush. Once the big stuff has been broken loose and rinsed away, the fine bristles on a floppy paintbrush will do a good job of completely cleaning this area, including the tight spaces on each side of the mount.

The front wheel suspension piece on this E-Type is dirty and in need of attention before the car can be entered into a concours event. Again, it is plainly visible when the front hood and fender unit are tilted forward. A rather stout, brass-bristled brush had to be used with Simple Green to remove grime that was embedded in the rough surface texture. Since the piece was neither painted nor polished, this kind of scrubbing had no ill effects.

an easy task, nor does it offer comfortable working positions for detailers. Therefore, serious detailers should consider raising the vehicle's front end, resting it on heavy-duty jack stands, and crawling underneath to clean the bottom of the engine, front-end structures, frame members, and everything else that cannot be reached from above. Once you start, you might wish you had spent more time cleaning this area with degreaser and a pressure washer.

Since you will be lying on your back and reaching up to clean, wear some sort of eye protection, like goggles. Not only will debris be falling from those items being cleaned, but your arms will most likely dislodge pieces of grit and road debris as you work. Consider wearing a dust mask or bandana as well, to prevent dirt particles from falling into your mouth and nose. While working under vehicles, most experienced detailers keep their head to the side of parts being cleaned to avoid bombardment by falling debris.

Cleaning Parts Removed from Engine Compartment

Once your engine and engine compartment are clean enough to eat off of, so to speak, turn your attention to those parts that were taken off earlier. Clean them with the same enthusiasm and thoroughness as displayed throughout the earlier phase of cleaning.

Your tools and cleaner are right at hand, and you should be able to use a wash bucket for items like the air cleaner housing, jacking equipment, windshield washer and radiator overflow containers, fan and shroud, radiator cap, valve cover oil filler cap, ignition wire holders, belts, breather tubes, air cleaner ducts, and so on.

Naturally, you cannot dunk a battery, alternator, or air conditioning compressor into a bucket of sudsy water, but you can do a thorough job of cleaning them with a damp cloth or wash mitt, toothbrush or plastic brush, and plenty of elbow grease. Use small brushes, or insert their handles into a folded cleaning cloth to

A small amount of dirt is located around the carburetor linkage assembly and along the gasket in front of this carburetor. The material must be removed, as it will surely be noticed by any judge and account for point deductions. Hose clamps need cleaning and should also be adjusted so that both are lined up evenly. Some touchup paint work is needed on the air cleaner box and along a support member.

Grease and dirt on this plastic container have to be removed. The task is most easily accomplished with the part out of the engine compartment. At the very least, it should be loosened from its bracket so that definitive cleaning with a paintbrush can be undertaken.

clean inside blade-like parts on alternators and into slots on parts that are too small for your fingers.

Overview

To some avid auto enthusiasts, such as Art Wentworth, spending a day or two (or more) simply cleaning everything under the hood of their special car or truck is both entertaining and rewarding. They will not strive to progress much further except to polish and wax chrome, brightwork, and painted surfaces, and maybe replace decals and stickers that were damaged or have seen better days. Their appreciation of automobiles rests with a vehicle's simple, clean, and natural state. For serious concours competitors, like George Ridderbusch, cleaning is a passion and any speck of dust, dirt, oil, or "unoriginal" debris is like dirt under the fingernails of a brain surgeon—it just has to go.

For those car buffs like Dan Mycon, with a fancy for bright hues and brilliant custom engine compart-

ments complete with lots of chrome or billet accessories and spring-colored wires, in-depth cleaning is simply a required step before all the *real* fun begins.

Regardless of one's ambition or intention for engine detailing, no vehicle will fulfill expectations unless all dirt, grease, and crud are completely removed from the motor and every part of the surrounding engine compartment. Paint will not stick to grease, dressing will not produce gloss on hoses caked with dirt, and a new set of billet valve covers will not look as good as they should when installed on a neglected intake manifold.

As I've said repeatedly, thorough cleaning is a critical part of engine detailing. If you are a conscientious auto enthusiast, this labor-intensive work should only have to be conducted one time on your car. Why? Because work of this caliber should make you want to follow a frequent, meticulous, and thorough cleaning schedule from here on out to prevent your engine compartment from ever getting that dirty again.

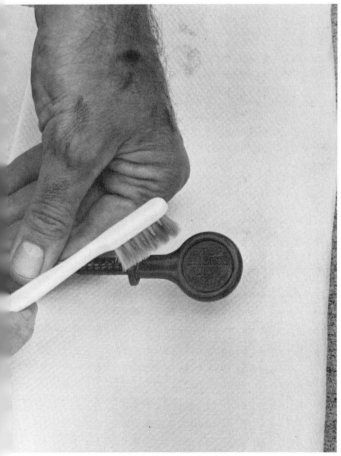

This pliable red oil dipstick handle is covered with a dirty film. Allowing it to remain in such a condition would just be asking for point deductions. All it took was some scrubbing with a soft toothbrush dampened with cleaner to make it look like new.

McKee uses a soft cotton towel to absorb water that has accumulated toward the rear of the engine compartment. As soon as cleaning has been completed, it is important to dry surfaces before stagnant water has a chance to dry in place and possibly cause spotting.

65

In this photo, McKee's E-Type looks ready for just about any concours competition. In reality, three more days will be needed to get it up to a seriously competitive level. A lot more time will be spent in the engine compartment looking for and removing dirt. Once all dirt has been wiped away, efforts will be focused on polishing and then paint touchup.

Chapter 5

Polishing and Dressing

Up to this point, attention has been focused on general cleaning and the removal of grease, dirt, and crud from the engine and all components located inside the engine compartment. Depending upon the condition of your engine compartment at the onset of this project, you should notice some degree of improvement, whether it be slight or something just shy of a miracle.

The task of bringing your engine compartment to a perfectly clean condition gets you to baseline. Every automobile, no matter its vintage or intended use, must be brought to this baseline before extensive detailing, customizing, or restoring work can begin.

It is like preparing a car body for paint. Once all of the dings have been taken out, sheet metal sanded and primed, and all masking chores completed, the vehicle has been brought to baseline. At that juncture, an autobody painter can spray on a car's factory-original color, a different stock tint, a custom metallic, pearl, candy, or an elaborate series of custom graphics, flames, or what have you. The choice is generally based on an automobile's vintage or classic status, its intended use, and the owner's imagination. But the car had to get to baseline first before anything else could be done.

The same holds true for engines and engine compartments. An owner like Art Wentworth, who prefers natural stock conditions, might be satisfied by simply continuing to polish and wax appropriate engine compartment parts and surfaces. Others, like Dan Mycon, may paint the engine, bead blast and paint certain assemblies, add on a few trick parts, and install neon-colored ignition wires. Concours buffs, like George Ridderbusch, will continue their detailing work until every nut, bolt, bracket, wire, hose, and so on in the engine compartment is painted, polished, shined, and conditioned to better-than-factory-new condition. About the only standard that must be met in engine and engine compartment detailing is the work needed to bring all parts and assemblies to baseline; how far you go after that is entirely up to you.

A bit of paint overspray and dirt needs to be removed from this support member before detailing work can proceed. Polishing will remove baked-on stains and other finish problems, but dressing will not do much to remove dirt that should have been wiped away before. Make a final inspection of your engine compartment to be certain everything is as clean as you want it to be, and then get ready for polish and dressing tasks.

Just a little more cleaning is needed before this super-charger and its components have been brought to baseline. After smudges of dirt are removed from the base of the carburetor, a detailer can begin polishing to bring the entire unit up to show car quality. A systematic polishing approach is necessary because of the configuration and distinct designs presented on all the various parts.

An outstanding engine compartment on this 1984 Corvette. Once it was brought to baseline, the detailer was able to meticulously dress all rubber and vinyl parts to bring out a high gloss. It is doubtful these parts looked this good when the car was first driven out of the factory.

Generic Improvements

Any engine compartment is a candidate for generic improvements. These include polishing and waxing paint on the hood underside, air cleaner, and inner fender aprons, applying dressing on hoses, arranging ignition wires in an orderly fashion along valve covers, and inserting them into their proper holders. Along with general efforts to make your engine compartment look tidy, consider doing a few preventive maintenance chores like cleaning battery terminals, tightening nuts, bolts, screws, and wire connections, checking vital fluids, and lubricating hinges and assorted linkages.

Polishing the paint on the underside of almost any hood will improve its finish and bring out a better shine. The hood on this 1966 Corvette is super clean, and polish work has enhanced the overall appearance of the engine compartment. Notice that the fender apron and radiator area have also been polished to exhibit a finish similar to the hood's.

Polishing and Waxing

Novice detailers are sometimes confused by the intended applications of polish and wax. In a nutshell, polishes are designed to clean and shine dull or oxidized paint, and wax is supposed to protect paint after it has been polished. Therefore, do not expect a wax product like Meguiar's #26 Professional Hi-Tech Yellow Wax or Eagle One's carnauba wax to bring out a high-gloss shine on an otherwise dull air cleaner, fender apron, or hood underside. You'll have to polish

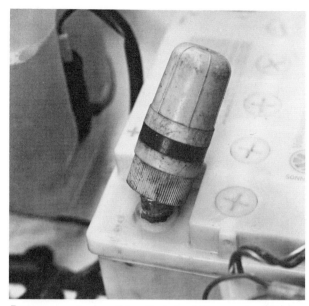

Battery care is something that should be ongoing. During an engine detail, though, be sure to brighten terminals and cable ends before putting it all back together after in-depth cleaning. Handy battery terminal cleaners are available at auto parts stores and they do a good job of cleaning both terminals and cable ends. To help minimize corrosive acidic build-ups on battery terminals, consider installing battery washers or using some other type of battery terminal maintenance product.

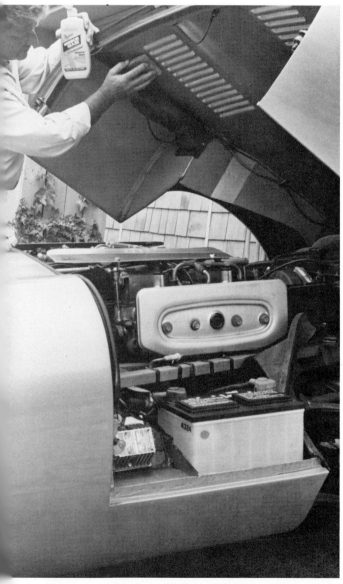

McKee uses Meguiar's #7 ShowCar Glaze to bring out a lively shine on the painted finishes of his car. Concours buffs like Ridderbusch and Hall use #7 exclusively and never apply wax over the finish. This is because a #7 shine will last long enough to endure concours judging, and they don't have to worry about the possibility of wax build-up or yellowing after extended periods. Polish courtesy Meguiar's, Inc.

Polish and wax are easily applied with soft, rectangular sponges like this one. The size is very manageable and its straight sides allow for controlled polish and wax application along grooves and next to rubber or vinyl materials. You can find this type of household sponge at most supermarkets or discount stores. Be sure to rinse sponges with clear water on a frequent basis to remove accumulations of dead paint and dirt residue.

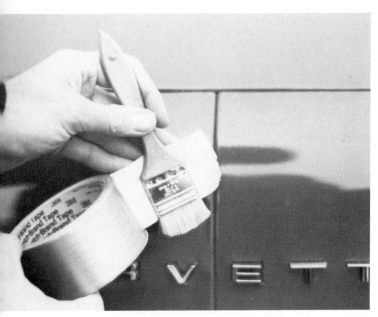

Novice detailers may think that a little wax or polish residue here or there looks cool and lets everyone know that they have polished and waxed their car. Well, avid auto enthusiasts know differently and prefer to remove every trace of wax or polish from every nook and cranny on their car or truck's body. A handy tool to help remove those dry polish and wax remnants is a cutoff paintbrush. The short, stout bristles do a great job of breaking up and whisking away powdery residue. Be sure to wrap a piece of heavy-duty duct tape around the metal paintbrush band to prevent metal-on-metal contact.

A cutoff paintbrush would do well to remove the dry polish and wax from the corner of this engine compartment. With the hood in position, access to this corner may be limited, but efforts should be undertaken anyway to remove this eyesore. Also, note the polish and wax build-up along the louvers on the cowling. A soft toothbrush might work well to remove residue.

those items first with a product like Meguiar's #7 Professional ShowCar Glaze. Once you have brought out a desired shine and luster, protect the finish with a quality carnauba wax product.

Since regular engine compartments are seldom exposed to direct sunlight and are also frequently coated with an extra-fine oily film that settles on surfaces from normal motor operation, severe paint oxidation is generally rare. However, if your vehicle's engine compartment paint work suffers unusual oxidation and dullness from being stored outdoors for a

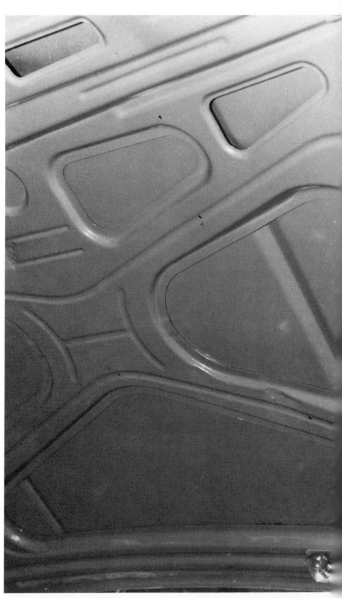

The underside of this hood looks like that of a show car. In fact, its host 1960 Chevrolet El Camino is a show vehicle. Notice the uniform shine on the piece. Finely polished and waxed surfaces do a much better job of repelling dirt and debris than unpolished ones. Maintaining a polished hood underside should make the task of wiping it down and keeping it clean a snap.

number of years, you may have to use a strong product like polishing compound to bring paint back to life. Terry Skiple likes to go over oxidized paint twice with polishing compound to remove dead paint surface layers, twice with a glaze like Meguiar's #7 to remove swirls, and then twice with very light coats of a quality carnauba paste wax to protect it all. He gets outstanding results.

A minor problem associated with polish and wax work around engine compartments is the amount of dust created when dry polish and wax is buffed off. To minimize dust while shining a hood underside, drape a sheet or large piece of plastic over the engine area to catch polish and wax particles before they fall onto the clean engine compartment. After completing all polish and wax work, use a soft, dry paintbrush to dust off parts, or a vacuum cleaner with a soft brush attachment.

For polishing and waxing the engine compartments of his cars, Wentworth uses Meguiar's Car Cleaner Wax, a one-step polish and wax product. It removes oxidation chemically and offers long-lasting protection. Wentworth likes it for two reasons: it shines and protects in one step, and when dry, it leaves behind very little powdery residue.

No matter which brand of polish or wax you use, application and removal must be done in an orderly fashion. Since engine compartment surfaces are

To match the hood, the firewall and fender apron on this 1960 El Camino are polished nicely and accented by a shiny air cleaner and tidy hoses and wires. All painted parts under the hood of special cars deserve polished shines and wax protection. Special in this context is not limited to show cars or concours candidates, but simply refers to any vehicle that is special to you.

It is not often one crawls under his or her car to clean, polish, and wax the oil pan. However, if yours is a special automobile that gets a lot of attention from admirers, you might consider spending a little polish and wax time in

that area. Once the paint finish has been polished to your satisfaction, a good coat of wax will protect it and also help you to keep it clean with little effort.

awkwardly shaped, try using a small, rectangular sponge for polish and wax application. Its small size and straight sides offer excellent maneuverability. Be sure to rinse sponges with clear water regularly to remove accumulations of dirt and dead paint. To prevent getting polish or wax on rubber, wrinkle paint, or other surfaces, outline affected areas with automotive masking tape. It is available at autobody paint and supply stores and comes in widths from ⅛in to a full 2in.

Buffing off dry polish or wax is best accomplished with a clean, soft cloth. Because of the confines inside engine compartments, smaller cloths work best. Cotton is preferred and old flannel shirts (with buttons removed), T-shirts, cloth baby diapers, or pieces of white flannel from a fabric store work well. Fold cloths into quarters so that you'll have a clean side to work with once one becomes soiled. When polishing and waxing are complete, buffing cloths and wash mitts can be thrown into the washer and dryer and used again. Wentworth says that the more he washes them, the softer they seem to get.

To remove polish and wax build-up from creases, seams, around bolts and screws, inside grooves and other tight spaces, use a soft, natural-bristled paintbrush. A 1in wide brush with the bristles cut to about

Spark plug wires are susceptible to accumulations of dirt that must be removed with cleaner. Once clean, Wentworth uses Armor All on a small applicator cloth to buff them to a deep, dark shine. Dull wires tend to make engine detail work look unfinished, whereas shiny wires enhance

engine pieces by offering a pleasant contrast in color. Be sure to apply dressing to applicator cloths first and then wipe wires. Spraying dressing directly onto wires causes unnecessary overspray on other engine parts.

Eagle One Mag & Chrome Polish does a good job of polishing brightwork, like the billet accessories pictured here. Use a clean, soft cloth as an applicator, and buff residue off with a clean section. If parts are polished outside of the engine compartment, hold them with a clean cloth during installation to avoid smearing surfaces with hand oils or other contaminants.

¾in works great. Wrap a couple of layers of duct tape around the brush's metal bands to cushion inadvertent bumps against painted objects and prevent heartbreaking paint chips. More stubborn build-up accumulations may require the use of a toothbrush or small detail brush. Whisk away dry polish and wax debris with a soft cloth or dry, floppy paintbrush.

The amount of polishing and waxing you do inside your car's engine compartment is limited only by your ambition. Naturally, the hood underside, fender aprons, and air cleaner are prime candidates, but what about the valve covers, radiator tank, painted sheet-metal supports at the front of the compartment, firewall, and large alternator brackets? Along with making those items look better, a coat of wax will help prevent dirt and grease from adhering to surfaces, making future cleaning efforts much easier.

Dressing Hoses and Rubber Parts

George Ridderbusch uses minimal amounts of Meguiar's #40 Vinyl/Leather/Rubber/Cleaner/Conditioner to bring out a perfect factory gloss on the hoses in his 1979 Porsche 928 engine compartment. Rather than spray gobs of dressing on hoses, he lightly mists a soft, clean cloth and then carefully buffs them until all excess is gone. Art Wentworth uses the same application technique, but prefers to use Armor All.

If you are concerned that a polish product like Simichrome or Eagle One may not produce the kind of results you expect, try a small portion on an inconspicuous spot first. Anodized finishes will not hold up under heavy polishing. In fact, harsh polishing could erase some anodized coatings. If such items need brightening, test a small amount of mild paint polish on an out-of-the-way spot.

Unpainted, chrome, or polished parts suffering baked-on stains may require something more than just plain polish. Here, #0000 steel wool is used to brighten a linkage rod. Place a small dab of polish on the steel wool and then apply. This technique works fast to bring out a shine and remove crud. A towel or rag should be placed under the part to catch falling steel wool fibers.

Polish and #0000 steel wool have made this aluminum support rod shine like it's supposed to. Neither a large amount of polish nor elbow grease were needed. It is best to rely on repeated applications, rather than chance scratches by forcibly and vigorously scrubbing pieces with steel wool. Do not overlook the bottom sides of parts like this, and consider taking small parts out of the engine compartment where they can be polished more completely and manageably.

Every detailer has his or her own product preferences and it is not so much the brand of dressing used that makes a difference as it is the application technique. Simply spraying dressing on and leaving it in place can cause hoses and other rubber parts to look unusually shiny, but could also cause them to attract accumulations of dust and dirt. Always work dressing into surfaces and buff off excess with a clean cloth to leave behind crisp, fresh, new-looking hoses and rubber parts.

For old and weathered hoses, splash shields, and other rubber parts, you might try bathing them in an

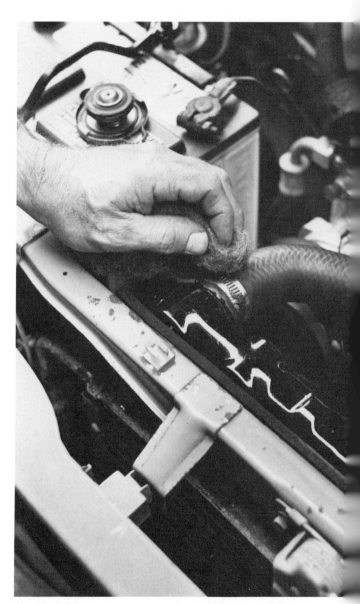

Unpainted hose clamps are quickly polished with #0000 steel wool. Regular chrome polish could also be used, but the polish residue left in clamp slots might be a little tough to remove. Steel wool does an equally good job of removing old paint overspray and crud build-up from clamps and other hose connections.

ample amount of dressing after they have been thoroughly cleaned. Like McKee, use a dedicated brush with soft bristles to work dressing into the material and all its crevices. Then, wash those parts with a soap and water solution to remove excess dressing and sheen. The outcome should leave parts looking like new. *Caution*: Never apply dressing of any kind to belts; the slippery nature of these products could cause them to slip over pulleys instead of gripping tightly.

Dan Mycon never uses a silicone-based dressing on his cars or engine compartment parts. This is because silicone causes freshly applied paint to fisheye, a condition that looks like several tiny volcanoes erupting. Because he owns and operates a professional autobody repair and paint shop, he cannot afford even the slightest hint of silicone dressing to infiltrate his work clothes or the facility's atmosphere. He opts instead to spend extra time cleaning hoses and rubber parts to utter perfection.

Polishing Chrome and Brightwork

Happich Simichrome Polish and Eagle One Mag & Chrome Polish both work well to shine dull or

This shield was designed to prevent water from splashing through hood louvers onto the valley of its V-12 engine. The water spot is an indication that the shield is doing its job. Spots like this are easily removed by buffing with a metal polish like Happich Simichrome or Eagle One Mag & Chrome Polish.

The shield is being removed so it can be polished to perfection out of the engine compartment. This also allows the detailer to clean and polish the underside of the shield, as well as the compartment area beneath the shield. It's also a good opportunity to clean the screws and washers that hold the shield in place. A toothbrush and cleaner will be used to dislodge any crud that has built up inside the screw slots.

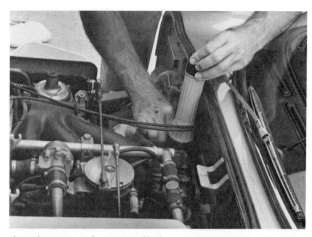

Anytime you take parts off of an engine, you must handle small screws, washers, nuts, and bolts with care. One of the screws used to secure the shield on this Jaguar got away from Wentworth and it took almost a half an hour to retrieve it. Place a can or tray close to your work area so small parts and fasteners can be immediately placed inside them, instead of on a cowling ledge or other spot.

lightly stained chrome and other unpainted parts. If you question whether such products will shine or will improperly mar highly polished alloy parts, try a small sample on an inconspicuous spot first. If it works, great. If not, try a much milder polish like Meguiar's #7 Professional ShowCar Glaze. A less abrasive material may require more time and elbow grease, but results may be worth it when scratches or swirls are prevented.

Chrome air cleaners, valve covers, timing chain covers, brackets, pulleys, bolt heads, and so on are objects that may require polishing. There is no need to use large amounts of polish, especially when parts are just slightly dull. A little polish goes a long way, so use it sparingly. Best results are often found by placing a small dab of polish on a soft, clean cloth and applying it with uniform pressure. Should chrome pieces exhibit small nicks or chips that are not big enough to warrant the cost of new chrome plating, try touching them up with Cold Galvanizing Compound for interim repairs.

Extra-dull parts may require two or three applications. Be sure to use a clean section of cloth with each maneuver so accumulated grit and debris are not rubbed onto adjacent surface areas. If simple polish applications are not strong enough to remove stubborn debris or the accumulation of minor rust deposits on chrome or other hardened polished parts, try using a dab of polish on a wad of #0000 steel wool. Once again, if you think the brightwork piece in question will not hold up under these vigorous polishing efforts, especially with steel wool, try it on an inconspicuous spot first.

Polishing small parts might be best accomplished with them off of the engine or out of the

engine compartment. Such parts would include ignition wire brackets, caps for the oil filler, brake fluid reservoir, and radiator, battery hold-downs, wing nuts, and the like. Polish as necessary and then apply a light coat of wax. Although high engine operating temperatures might dissipate wax coats quickly, the protection offered in the meantime is good. Prevent fingerprints and other oil smears on freshly polished and waxed parts by holding them with a clean cloth during replacement.

Efforts to polish and shine tubing lines will ultimately result in polish build-up around connectors. Use a cutoff paintbrush to whisk away dry residue. Hold a cloth under such items to prevent powdery particles from landing on nearby surfaces.

Small recesses requiring polish work can be difficult to reach. Innovative and meticulous detailers have painstakingly used cotton swabs to both apply and buff off polish in such areas. Thin grooves along custom chrome valve covers, slots on bright fins, and holes in billet pulley assemblies are typical cotton swab candidates. If you have trouble fitting a fold of cloth into similar spaces for polishing and buffing, try using cotton swabs or the wooden end of an artist's paintbrush inside a cloth. Never use a screwdriver or other hard metal object. The chances for that sort of tool poking through cloth fabric are great, as are the chances for inflicting unnecessary scratches or gouges on the brightwork piece you are working on.

Overview

The engine compartment detailing techniques described in this chapter are some that have worked successfully for many avid auto enthusiasts and professional detailers. They are not intended to be the last word in detailing, however, and you are encouraged to try different techniques as you see fit.

Polishing and waxing are important tasks that should never be taken lightly. The appearance of virtually every painted surface in an engine compartment can benefit from a bit of polish and wax, and this will go a long way toward protecting the paint finish as well. As you dive deeper and deeper into your detail project, don't be surprised if the day passes by in a flash. Many serious enthusiasts schedule entire three-day weekends for nothing but engine detailing, and frequently find that three days are not enough.

You see, as one part of an engine compartment perks up and stands out, others may start to look worse and worse. Before long, you might find yourself sticking bolts and screws into pieces of cardboard so you can paint them, pulling apart grille sections to clean areas underneath, dismantling fan shrouds for cleaning and to gain access to the radiator to straighten its fins, using an artist's paintbrush to touch up paint stripes inside valve cover grooves, and turning your attention to other intricate details that winning concours veterans like Ridderbusch simply take for granted.

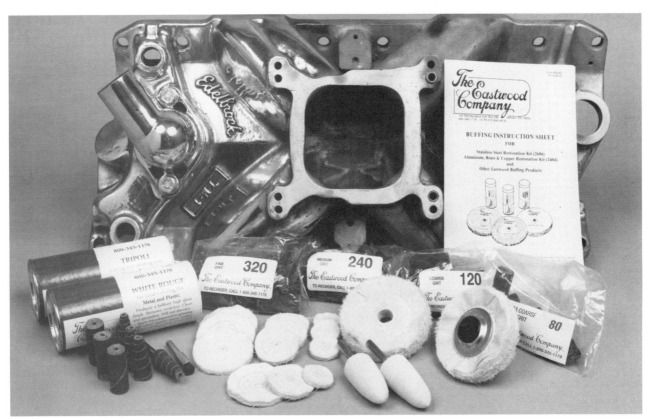

Rough cast-aluminum manifolds can be made to shine like chrome when polished with accessories included in this Aluminum Manifold Kit available from The Eastwood Company. Everything you'll need is right here, except a grinder or drill. Be sure to follow the detailed instructions. Photo courtesy The Eastwood Company

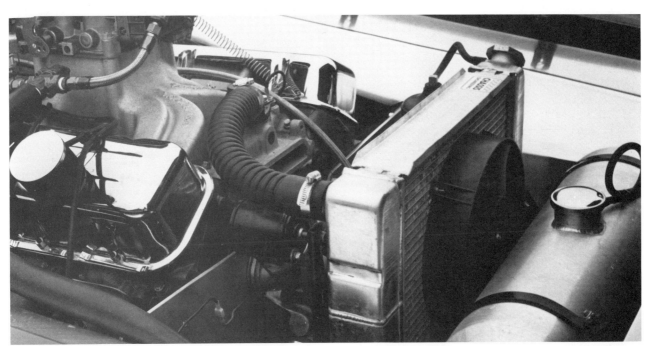

The engine in this race car sure looks good with polished valve covers, bright hose clamps, and a clean overall appearance. To reach tight places too small for your fingers, insert the wood handle of an artist's paintbrush into a cloth and use it for polishing and buffing. Never use a screwdriver or other hard object. They can easily stick through the fabric and cause deep scratches on otherwise perfectly smooth finishes.

A Saturday morning was spent cleaning this Chevy V-8 engine compartment, and the afternoon was used to polish paint and brightwork and dress hoses and wires.

The dark material running along the top of the heater box is factory sealant, not grease or dirt. What a difference a day can make!

Chapter 6

Paint Work

The best way to achieve a perfect paint job on the block, heads, intake manifold, and other engine components is to pull the engine out of the vehicle and secure it to an engine stand. Out in the open, an engine is totally accessible for required first-class sandblast work, deburring, intricate masking, epoxy primer application, and quality spray painting. An engine slated for a complete paint job, especially one that will receive a custom urethane color change, must be yanked out, stripped of accessories, and put on a stand.

Painting the Engine

For general touchup or meticulous detailing, you can successfully paint most engine parts while they remain intact in the engine compartment. Have two cans of engine paint available if painting with aerosols, or at least a full quart if doing the job with a regular spray paint gun. Maximum quality and coverage depend on your ability to see and reach various areas around the engine. So, plan to remove obstacles like the alternator, air conditioning compressor, and belts as necessary.

Other items such as ignition wires, carburetor throttle return springs, and heater hoses can be held out of the way with your free hand as paint is sprayed near them. If you so desire, remove valve covers and other parts to paint them separately. Although everything in the engine compartment may have been carefully and thoroughly cleaned, wipe parts off with a wax and grease remover before actually spraying paint. This type of mild solvent will remove traces of cleaner film and other materials that could adversely affect paint finishes. As needed for rough spots, especially on otherwise smooth surfaces like valve covers, use 150 grit sandpaper and a sanding block.

Engines normally get hot while they run and therefore require paint products that will stand up to high temperatures. Hence, the availability of high-temperature engine paints at auto parts stores and autobody paint and supply outlets. Stock colors are generally limited to such colors as Chevrolet Orange, Ford Blue, Chrysler Blue, and so on. Be sure to read

A first-class paint job requires that the engine be pulled out of the compartment and placed on a sturdy engine stand. Out in the open, a detailer is able to clean, grind, or sandblast those parts of the block that need it, accomplish intricate masking tasks, apply recommended coats of epoxy primer, and then accurately spray on coats of paint. The same kind of detailed preparation and paint work can be afforded those parts that were taken off the engine.

An abrasive disc can be used to remove paint, scale, and rust from an engine block. It is intended for work on smooth surfaces to speeds of no more than 5000rpm. The die grinder, in this case, will be operated at only partial capacity. This combination, used at relatively slow speeds, will help to remove the bulk of flaking or peeling paint from the engine block. Used with a regular drill motor, the abrasive disc will strip valve covers and other smooth parts. Tools courtesy The Eastwood Company

paint can labels to be certain the color you have selected is correct for your year, make, and model of vehicle; some basic colors, like blue, may come in slightly different shades. The Eastwood Company offers a range of engine paints which includes but is not limited to Chevrolet Blue, 1951-62 six-cylinders; Ford Green, 1929-40, all models; Ford Dark Blue, 1941-48; Ford Light Blue, mid-1960s and newer; and Detroit Diesel Green. These paints sell for around $4.95 per spray can.

Engine paint in stock colors is also available in quart cans at autobody paint and supply outlets for those detailers who prefer to use small detail spray

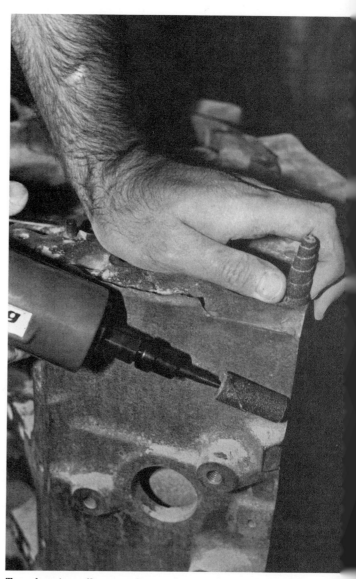

Two abrasive rolls, part of an engine porting kit. Designed to be used with a high-speed die grinder, the rolls smooth rough head castings when blueprinting an engine. They can also be used to smooth rough parts of engine block exteriors in preparation for paint work. Rolls courtesy The Eastwood Company

Gasket cleaners are used on the end of a drill motor for removing gasket residue and other baked-on materials. They are perfect for detailers restoring and rebuilding engines where engine decks, head surfaces, oil pan rails, and other surfaces need to be brightly cleaned in preparation for new gaskets or paint. Cleaners courtesy The Eastwood Company

paint guns, in lieu of spray cans. When working with a detail paint gun or regular paint gun, have a large piece of cardboard handy to use for testing spray patterns. In tight spots, you'll need a tight spray pattern and it is much easier to adjust pattern controls against a piece of cardboard than it is on an engine block.

Before painting, take some time to mask off certain parts of your engine. Use automotive masking tape which is available in a variety of widths at auto paint stores. It will not leave behind glue residue, and it is also designed to prevent paint from seeping underneath the edges, unlike some cheaper generic brands. Mask along protruding head and other gaskets so their stock color will remain intact. Use tape to cover water temperature and oil pressure sending units, chrome water pump housings, radiator and heater hose ends and their clamps. As necessary, use thin tape strips for slender items like gaskets, and wide tape for larger items such as exhaust manifolds and carburetor bases.

Pieces of aluminum foil, large towels, and sheets of masking paper work well to cover engine compartment assemblies not slated for paint. Place them over inner fender areas, on top of frame members, or drape them down in front of the firewall. Use masking tape to secure.

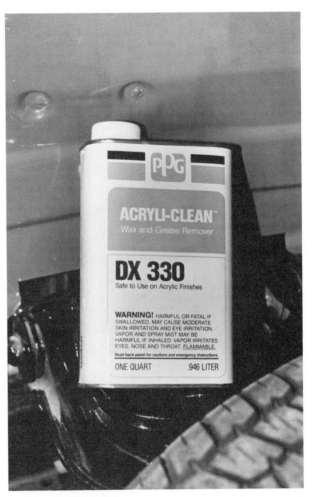

Empty engine compartments should also receive plenty of prep work. In-depth cleaning, of course, is mandatory, but along with that you should repair cracked seam sealer applications, pull out and fill dents, sand rough finishes, and remove and repair all signs of rust or corrosion. This engine compartment has received some bodywork repair and will be meticulously cleaned before painting.

Once all of the preparation tasks have been completed and you are ready to paint, take a few minutes to wipe off surfaces with a wax and grease remover. Mild solvents, like PPG's Acryli-Clean, do an excellent job of removing lingering soap and cleaner films, as well as any dirt or grease particles that may still be on the surfaces. All professional automotive painters clean their work surfaces with a wax and grease remover before spraying paint.

Another handy item to have is a paint block. The bottom of a shoe box or other piece of lightweight cardboard is ideal. Position the paint block between the part to be painted and those behind it. A paint block can be wedged in place or held steady with your free hand. The thin nature of lightweight cardboard allows you to fold or bend it into a variety of shapes as required. Whatever means you employ to block paint overspray, be certain the material is positioned correctly and held firmly in place. Efforts to maximize protection will be greatly rewarded when it comes time to check for and remove overspray.

Engine paint will run just like any other paint product. Therefore, it is best to apply two or three light coats as opposed to a single heavy one. Results are best when paint is applied in warm weather. Consider warming paint cans in a sink full of lukewarm water, a temperature that is *not* too hot for your hand. This will help paint pigments and binders to mix better and will maximize the propellant's capability. If you use a detail spray paint gun, be sure your air pressure regulator is properly adjusted and that all water condensation has been drained from the air pressure storage tank. Optimally, you should have an inline air dryer or water trap connected to your air source.

Start painting an engine at the back of its intake manifold. This way, you can rest your free arm on top of a nearby valve cover. Hold the paint can so that your index finger is in line with the nozzle opening; wherever your finger points should be the direction in which paint sprays. Quick back-and-forth spurts, as opposed to long, continuous sprays, should reduce the possibility of paint runs. Work your way around the intake manifold using your free hand to hold wires and other objects up and out of the way. Pull off the carburetor throttle's return spring to allow access to that side of the intake manifold.

Most auto parts stores carry ample supplies of high-temperature engine paint in a variety of stock colors. If you plan to paint your engine with an aerosol, be sure cans purchased are from the same paint lot. This is indicated by a batch number on the bottom of the can. Identical numbers should guarantee that paint in both cans are an identical shade and tint because they came from the same batch of paint.

If you decide to paint your engine with a detail spray gun or regular spray paint gun, you will need to thoroughly clean it after painting. Never use a hard object, like a screwdriver, to clean a paint gun. Instead, use a quality brush designed for cleaning the internal passageway and exterior portions of such tools. Brushes like these will do a good job of cleaning without risking scratches or other damage to gun components. Brushes courtesy The Eastwood Company

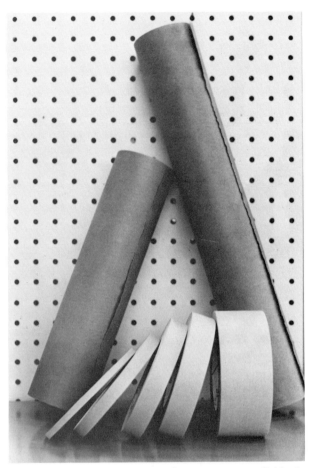

Automotive masking tape and paper are available in various sizes at autobody paint and supply stores. This selection of masking tape ranges in widths of 1/8in to a full 2in. It will not leave behind a glue residue when pulled off, nor will it allow paint to penetrate edges. Automotive masking paper is designed to resist paint penetration and is much preferred over rather porous and flimsy newspaper.

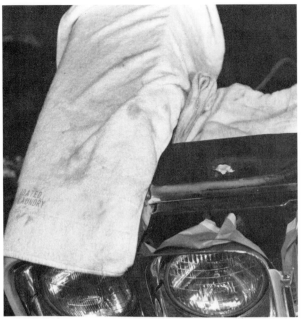

Preparing engine compartment surfaces for painting may require a bit of innovative masking. Here, large towels are used to protect a fender, and automotive-grade masking paper and tape cover grille pieces in preparation for painting a cross-member. Note that the hood bumper is masked and the radiator cap has been removed. The next step will include thorough cleaning with a wax and grease remover.

A thin, lightweight piece of cardboard is used as a handy paint block. Detailers hold paint blocks with their free hand while painting with the other. Placed between the part to be painted and those behind it, paint blocks do a great job of preventing overspray on items already painted or those not scheduled for any paint work. The bottom of a shoe box makes an excellent paint block.

The air hose connected to the detail spray gun should not be dragged over radiator parts, fenders, or other components. Drape it over your shoulder and across your back to keep it out of the way. You could fish an old piece of electrical wire (one with a plastic coating) through an opening on the underside of the hood, tie it in a loop, and use it to hold your air hose up and out of the way. Slack in the hose could be pulled through as needed to reach engine areas close to the firewall or down lower in the engine compartment.

Once the intake manifold is completed, paint the front of the engine block. In some cases, access is best provided to lower front engine block areas from underneath. To reach tight spots, you may have to turn a paint can upside down. Since paint can feed tubes extend close to the bottom of their containers, you might have better luck using full cans in this position.

Water or contaminants in compressed air lines will wreak havoc with spray gun paint finishes, as well as sandblasting maneuvers and other pneumatic operations. A moisture separator, like this one from Eastwood, is designed to remove water and dirt that may collect inside compressed air lines. Water traps should be connected to an air outlet and hoses plugged directly into them for operation. This way, water and dirt are filtered out of all permanent compressed air piping before air reaches your hose. Photo courtesy The Eastwood Company

Reaching spaces around the front of the engine block is not always easy. Belts are generally in the way and so are the power steering unit, brackets, and so on. Consider removing such obstacles when attempting to paint. In addition, aerosol paint cans may have to be used upside down in order to get spray nozzles close enough to lower engine sections. Have a full paint can set aside for this job because full cans will spray much better in this position than half-full ones.

To accurately spray engine paint on the head, spark plug wires have to be removed. Do this one by one so their firing order is not mixed up. To protect spark plugs from paint overspray, you can remove them and insert Spark Plug Plugs or a set of old plugs that can be discarded after painting. Spark Plug Plugs might also be handy for chores involving sandblasting, grinding, or other cleaning work. Plugs courtesy The Eastwood Company

A small section of garden hose has been placed over a spark plug to protect it against paint overspray. This innovative idea works well during quick paint touchups. When you are faced with a problem, think about different ways of achieving satisfactory results and maybe you'll invent some shortcuts that work just as well as other tried-and-true methods.

Painting the sides of an engine may be made difficult due to obstructions such as spark plugs, ignition wires, exhaust manifold, or header. Ignition wires and spark plugs are easily removed. Write down the position of each ignition wire on separate strips of masking tape and then attach those strips to each wire so you'll be able to tell what goes where when you put them back. To avoid getting paint all over the spark plugs, pull them out and insert Spark Plug Plugs into their head openings; they are available from The Eastwood Company. Or, put in a set of old spark plugs that won't be damaged by paint overspray. Strips of masking tape will protect the exhaust manifold and header.

For spray painting engine areas behind header pipes, you might try attaching a nozzle tip and spray tube from a can of WD-40 lubricant to your can of spray paint. Spray some paint out of the tube toward an open area first to blow away any WD-40 material that may be lingering inside the nozzle or tube. Then go to it. This is a handy technique used by many professional detailers and avid auto enthusiasts. Along with being able to direct paint spray through openings in header pipes, you can also use a WD-40 nozzle and spray tube to paint inside other tight spots,

Another innovative tip developed by a meticulous engine detailer is to use a WD-40 nozzle and spray tube on your can of engine paint. The spray tube works great for directing fine paint sprays into tight spaces like behind header pipes, between belts, and so on. The tubes do not generally fit on regular spray can nozzles, so be prepared to pull the entire nozzle off of the can of WD-40 and push it into your paint can. Be careful not to spray paint in your face as the nozzle is pushed onto the paint can.

Applying engine paint on lower block sections may be difficult with accessory parts in the way. You may get a better working angle from below. Removing parts is another option that will help to make paint work go more smoothly. The WD-40 spray tube can come in handy here as well. Consider masking off parts to prevent overspray.

Paint work around the engine compartment can be limited to touchup, or involve a full-blown repaint, or anything in between. The paint chips in this area need to be sanded smooth and then treated with a chemical rust remover, like Oxi-Solv. Epoxy primer should then be applied, followed by coats of color-matched paint. Paint blemishes such as this need to be repaired early on before rust has an opportunity to start and spread.

like behind the power steering unit, on top of the bell housing, and anywhere else you see fit.

Use a drop light to check your progress. Apply a second or third coat as needed. Touch up spots that were missed and make a note of any overspray problems that must be corrected later. Satisfied that the job is finished, turn your can of spray paint over and hold the nozzle open until propellent *only* (no paint)

is discharged. This should clear the nozzle and keep its passageway open for future use. If you used a detail spray gun, now is the time to empty remaining paint back into its container and clean the gun thoroughly.

Painting Other Engine Parts

Besides the motor, a number of other engine compartment items may look dull and need a fresh

Any number of engine compartment pieces may need a coat or two of paint. The paint on this bracket is flaked and needs to be sanded before being touched up. The best paint jobs are accomplished when parts are out of the engine compartment. However, definitive masking and careful paint spraying can be accomplished with parts in place to achieve satisfactory results. Use towels, masking paper, and tape to block off surrounding areas to avoid unsightly paint overspray.

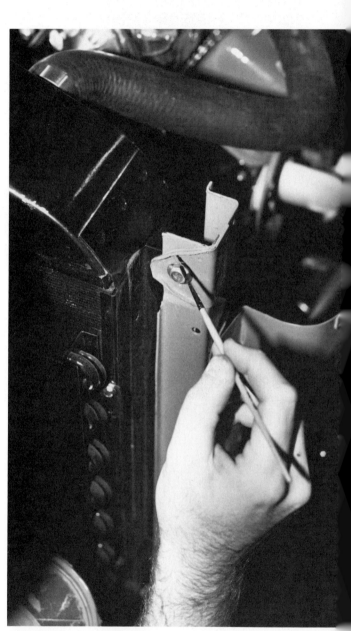

Dan Mycon is using an artist's paintbrush to touch up paint on the top of a radiator support in the engine compartment of his 1948 Chevy. These brushes are available at artist supply stores, sign painter supply outlets, and sources such as The Eastwood Company. After bristles have been cleaned and dried, coat them with petroleum jelly or motor oil to keep them properly shaped during storage.

coat of paint. Although concours cars and show vehicles will require a great deal more preparation, attention to detail, and definitive work, you can complete a number of engine compartment painting tasks on regular drivers with motors in place and simple common-sense techniques. Just remember, if you want absolute perfection on every engine compartment component, you will have to pull the motor, remove everything from the engine compartment, properly prep, sand, and mask, and then paint it all in a professional paint booth. In lieu of that method, and

the costs and labor involved, take your time and be patient in preparing for and applying paint around your engine compartment while the motor is still in it.

Painting a firewall and inner fenders should be limited to touching up paint chips, nicks, or scratches on any color other than black. Chips in factory body colors can be repaired by using bottles of touchup paint available at auto parts stores and autobody paint and supply outlets. Bottle caps are generally equipped with a small applicator brush. You can also

This is a Chip Kit available through Pro Motorcar Products, Inc. It contains all the tools necessary to achieve perfect paint chip repairs. The kit includes, among other things, clean paper matchsticks, a favorite paint applicator used for years by enthusiasts to repair nicks and chips. It also includes the Eliminator, which is a hand-held tool

that gently removes rust and other contaminants from metal by way of super-thin fiberglass strands bonded together like a pencil lead. It works great for cleaning and prepping paint chips before paint application. Kit courtesy Pro Motorcar Products, Inc.

Concours competitors and restorers have to be concerned about which paint colors are applied to their engine compartment parts. Paint shades must be in line with factory-original standards. Frame and suspension components on many automobiles were originally black but had a degree of gloss somewhere between gloss and flat. Eastwood's Chassis Black has been designed to meet the gloss standards established by factory paint codes for frames and suspensions. Photo courtesy The Eastwood Company

A lightweight cardboard paint block is used while black paint is applied to the top of a radiator tank. Masking tape has been placed along the blue cross-member in front of the radiator, the sticker on the fan shroud, the radiator hose end, and its clamp. This is a quick way to touch up paint without having to dismantle parts.

The paint work on this part is heavily orange peeled, evidenced by the rather rough texture finish. It could have been caused by contaminants on the surface that were not cleaned off before painting, or the application of second and third paint coats before preceding coats were allowed to dry. Take engine compartment painting seriously and be sure to follow instructions and recommendations provided on paint can labels or informational sheets available at autobody paint and supply stores.

Engine paint overspray on hoses, wires, and other parts is an obvious sign that engine detailing was done haphazardly or by an inexperienced detailer. All overspray must be removed in order to make detailing efforts look professional. A small amount of lacquer thinner on a clean cloth will quickly remove overspray from wires, hoses, carburetor base, and vacuum unit. Overspray is evidenced on these electrical connections. It makes them look dirty and worn. A small wad of #0000 steel wool may have to be used to scrape off encrusted old paint.

use an artist's fine paintbrush or the clean end of a paper matchstick. The Eastwood Company and Pro Motorcar Products, Incorporated, offer special automotive paint touchup kits. They include all the tools you'll need; all you have to supply is the paint.

Some of the more typical engine compartment items you may choose to paint are frame members, steering boxes and linkages, brake fluid reservoirs, radiators, power steering units, battery trays,

brackets, and mounts. Many novices spray gloss black because they like the shiny finish it produces. Other, more serious detailers opt for semi-gloss black, or buy original tint colors at an autobody paint and supply store and apply them with a detail spray gun.

Shop around at auto parts stores and auto paint outlets for paint products designed especially for engine compartments. The April 1991 issue of *Car Craft* magazine ran an article titled, "101 Paint Tips." A formula for semi-flat black paint for GM chassis was included as follows: "3 quarts of mixing black and one quart of flattening agent. Use PPG DTR601 quick-dry reducer." The article suggests using enamel on frame and suspension pieces because it is more durable than lacquer and will quickly wipe down for show purposes.

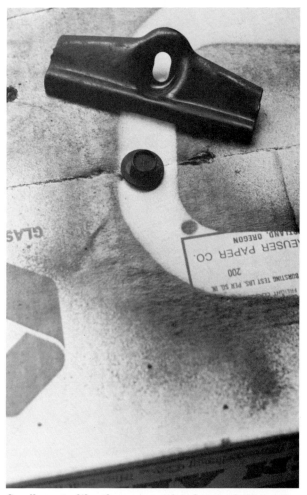

An impatient engine detailer sprayed black paint in the area behind this grille. It might look presentable from a distance, but not close up. You can see the orange peel finish on black parts and the overspray on brightwork. This grille should have been removed to accomplish quality painting. Overspray can be removed with #0000 steel wool. As far as paint finish is concerned, parts should be sanded smooth, properly prepared, and then painted again.

Small parts like these are painted out of the engine compartment on top of masking paper or cardboard. Bolts and screws are inserted into cardboard so only their heads are painted, not their threads. This battery hold-down and accompanying bolt will be cleaned, sanded, wiped down with wax and grease remover, and then painted with Chassis Black.

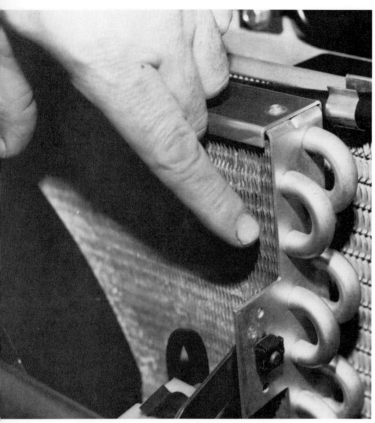

The thin edge of this bracket on Ridderbusch's Porsche 928 was starting to show a hint of rust. It was sanded smooth and brightened, touched with a chemical rust inhibitor, and then painted with a color that matched the rest of the bracket. The job was done correctly and there is no evidence that any problem ever existed. Engine detailers have successfully used Krylon Dull Aluminum paint to touch up and match the finish on cadmium-plated parts.

A great many used cars sitting on dealers' lots have had their engines detailed. To make them sparkle like new, detailers spray entire engine compartments with Crystal Clear Lacquer paint. This paint makes hoses, wires, painted surfaces, and otherwise dirt-stained surfaces look extra-shiny and new. The problem is that this condition does not last very long. As seen here, clear paint yellows after a short time to make engine compartments look ugly. Never spray clear lacquer on engine compartments that are in good to excellent condition and never use it on special vintage, classic, or custom cars. Save this quick trick for sleds that you might be detailing in preparation for their sale.

The Eastwood Company offers Chassis Black paint which is produced with a specific percentage of flattening agent that develops just the right degree of gloss to make parts look factory-original. It costs around $9.95 per can, with discounts for three or more. In addition, for about $4.50 each, they offer Spray Gray for cast-steel parts, Detail Gray for stamped and machined steel parts, and Aluma Blast for cast-aluminum parts.

Obtaining a clear shot at some engine compartment parts with a paint can is not always easy. If possible, remove items so they can be painted out of the compartment. For others, use masking tape, paper, rags, and a paint block as best you can. Although you may certainly use any color desired, many engine components such as the power steering unit, alternator brackets, and the like look great painted black. Others, like the brake fluid reservoir cap, should be painted either bright silver or gold, whichever you prefer.

Apply paint as you did to the engine, with uniform back-and-forth movements and short bursts of spray. Prolonged applications will result in runs, so opt for two or three light coats as opposed to a single heavy one. Use a drop light to check your work and don't be afraid to stop and add strips of masking tape, as needed, for protection against overspray.

Removing Overspray

One of the more obvious marks of an inexperienced engine detailer is paint overspray on engine

Battery trays can deteriorate from the effects of acidic residue that has dripped down onto them. In-depth engine compartment cleaning should include the removal of the battery and the subsequent cleaning of the tray. They can be coated with regular paint after being properly sanded and prepared. Or, you may consider using a special battery tray coating in lieu of regular paint. The Eastwood Company offers such a product that dries to a satin-smooth finish and also resists the harmful effects of battery acid.

parts. Chevrolet Orange on a power steering unit or Ford Dark Blue on carburetor parts looks amateurish, sloppy, and detracts from an otherwise tidy engine compartment. Actually removing overspray is easy; the tough part is finding every overspray blemish, no matter how small it may be.

Clear paint is not good for engine compartments, nor is it recommended for bare-metal parts that could rust under coatings and then present ugly, stained colors. Aluminum is different, however; once it is polished, it can retain its shine with a coat of Eastwood's Polished Aluminum Clear Coat, which can be brushed or sprayed on. This may be a good alternative to weekly aluminum polishing. Coating courtesy The Eastwood Company

Removing overspray from wires, linkages, hoses, clamps, chrome parts, and so on simply requires a dab of lacquer thinner on a clean cloth and some *gentle* rubbing. Fresh paint overspray will quickly disappear with a minimal amount of effort. Look for overspray on carburetor parts, radiator and heater hoses, ignition wires, electrical connections, water temperature and oil pressure sending units, belts, pulleys, and so on. Blemishes on a rough-textured unpainted exhaust manifold may not completely disappear with lacquer thinner and a rag, but should quickly burn off later, after the engine has operated for a while.

Although they should have been covered with masking tape, do not use lacquer thinner to remove overspray from decals, stickers, or emblems. As stated earlier, its potent nature will quickly remove any overspray, of course, but will erase the printed material along with it. Instead, carefully use a wax and grease remover or paint thinner. They are much less potent and are more apt to remove overspray without damaging anything else. Use a light touch,

Stainless Steel Coating works great to keep parts like exhaust manifolds and headers looking as if they have just been sandblasted. This product dries to an attractive metallic satin finish. It is designed to resist rust corrosion, and is unaffected by the hot and cold cycles of normal engine operations. Parts should be sandblasted or sanded before application. Coating courtesy The Eastwood Company

and avoid contact with the emblem or decal as much as possible.

Removing black overspray from a blue or orange engine block may also result in blemishes on the engine. Should that occur, simply touch up those spots with an appropriate engine paint. A light touch generally removes overspray with few problems but since lacquer thinner is such a strong solvent, underlying paint finishes are frequently affected.

Painting Accessory Parts

A host of small items should be painted while out of the engine compartment. These can range from air cleaner wing nuts to battery trays. Before painting, make sure they are clean and then wipe them off with a wax and grease remover. Some parts, like wing nuts and radiator caps, can be placed on a piece of cardboard and painted. Others, like air cleaners and valve covers, might turn out best if they are suspended by wires from the edge of a garage door or rafter. This allows for complete paint coverage on all sides, tops and bottoms included.

As with other engine compartment parts, paint color is important. Stock black air cleaners might look best painted with a black semi-gloss or with the same paint used for other black parts, like The Eastwood Company's Chassis Black. The same goes for black oil filler caps, voltage regulator covers,

brackets, fans, and so on. Parts that came stock with plastic coatings can be rejuvenated by using plastic parts dip products, available in various colors through autobody paint and supply stores and suppliers like The Eastwood Company.

Bright silver makes wing nuts and older radiator caps look new again. Krylon Dull Aluminum spray paint is said to be a good color match for cadmium-plated parts that are chipped but not quite ready for complete replating. Depending on the style and condition, you can also use bright silver to dress up alternator housings and pulleys. Unless yours is a concours contender or simply a vehicle you wish to remain factory-original, you can paint items black, silver, or any other color you desire—although, straying too far from stock color options or typical street rod or muscle car setups might detract from an engine compartment's appearance rather than enhance it.

Battery trays are highly susceptible to corrosion caused by acidic products that either leak out from around battery caps or seep down from dirty terminals. Regular black paint does not generally hold up well under those acidic conditions and you may get better and longer lasting results using a product designed for battery trays, like Eastwood's Battery Tray Coating. This heavy-duty, flexible coating looks like paint but resists the harmful effects of battery acid more effectively. It dries to a smooth, satin finish and can be cleaned with regular soap and water. It is recommended that trays be primed before coating.

Bare-metal parts should not be coated with clear paint because they could rust underneath and discolor. But, aluminum parts that have been bead blasted may continue to look good for a long time if painted with an aluminum clear coating. Clear may also work well to protect the finish on painted bolts. In addition, *Car Craft*'s "101 Paint Tips" article suggests baking bolts that are dry to the touch, after painting them flat black, on a cookie sheet at 325deg Fahrenheit to achieve the special black oxide finish featured on many original muscle cars.

Exhaust manifolds should be sandblasted and then washed with lacquer thinner before being painted with Eastwood's Stainless Steel Paint. Other items like brake fluid reservoirs and steering boxes should be bead blasted in a cabinet and then coated with a product such as Eastwood's Cast Gray Iron Paint or a semi-gloss black. For aluminum parts, try Eastwood's Aluma Blast. Headers can be painted, chromed, or coated with special materials. They must be bead blasted and thoroughly cleaned first, however. Autobody paint and supply stores, some auto parts houses, and mail-order companies like The Eastwood Company and Bill Hirsch, carry extra-high-temperature coatings for headers. Be sure headers fit properly before painting them.

Chapter 7

Overall Appearance Options

Automobiles were originally manufactured to provide transportation that would be more comfortable, reliable, and faster than four-legged animals. That basic intention has held true, obviously, but automobiles have become much more than simple transportation for many people. Besides offering a means to get from point A to point B, all types of vehicles are now able to carry heavy payloads, be operated in all kinds of weather, in on-road and off-road conditions, be raced against each other in all sorts of events, and

With the hood closed, you might never suspect that this heavily chromed and perfectly detailed supercharged engine powers a 1972 Chevrolet Blazer. The rest of the vehicle looks great, but the engine compartment usually knocks admirers for a loop. Engine compartments are a *natural point of interest for all sorts of auto enthusiasts. They can be modified in almost any way an owner desires. The key is to make the engine and all its accessories look like a matched set.*

can even be set up to look different from other vehicles of the same make or model. It is this desire to make one's vehicle stand apart from others that fires a detailer's ambition.

Given the time, expertise, and financial support required, you can alter the appearance of your special car or truck to resemble almost anything you desire. Of course, there are strict governmental requirements that must be met in order to keep vehicles street legal and safe, but lots of appearance options have been successfully incorporated into lots of automobiles for the joy and satisfaction of many a maverick automobile owner and customizer.

A special vehicle's exterior paint and attire, and interior upholstery and trim command a great deal of attention from admirers. However, once the hood of that vehicle is opened to expose a meticulously prepared engine compartment, admirers flock to it and generally rave over its appearance. This is evidenced at car shows when you see more people peeking into the engine compartment of a winning entry than you see drooling over its big tires or pretty carpet.

Engine compartments are set up any number of different ways, depending upon their functions and their owners' preference. Factories build engines and their accessories to provide users with dependable power and comfort options. Engine compartment designs must include equipment to provide horsepower and a means to operate lots of other items like power steering units, brakes, electrical parts, air conditioning, heaters, and so on. Engineers have to be concerned about all those things fitting into an engine compartment to start with, and then about their ability to function both individually and in unison.

It would not be easy to modify this engine compartment because factory engineers have already packed it full with a V-12 engine and lots of other equipment. Some factory parts are designed to look good as well as function correctly, but cost factors generally prevent manufacturers from adding polished, chrome, and other custom-like accessories. This Jaguar is probably best off maintained in its factory-original condition.

Hot rod sporting a traditional flashy engine and engine compartment. Hot rods are supposed to look this way, and any admirer would be disappointed to see anything different. A lot of work and money has been put into this rod and the owner has spent considerable time maintaining it. Some custom accessory options here include a polished firewall and intake manifold, chrome valve covers, headers, and brake fluid reservoir, stainless steel braided hose and wire covers, anodized hose ends, and neon colored spark plug wires.

Although engine compartment appearance is important to factory engineers, it rightfully takes a back seat to operating efficiency and economy. On the other hand, private owners are not handicapped by having to provide dependable and economical engines and accessories for assembly lines full of cars. They are able to focus their attention on just one vehicle—their own. If they like wiring in custom colors and engine parts with lots of chrome or billet goodies, they are certainly free to install them. The only limits to their customizing efforts are keeping the vehicle street legal, ensuring that aftermarket parts are compatible with existing ones, and working within their budget.

Engine compartment appearance options can provide a number of images ranging from a simple, clean, and natural factory-stock look to Concours d'Elegance-winning beauty to muscle-car high-performance machismo to street rod glitz. The choice to maintain a stock engine compartment or go the custom route rests with the individual. Although consideration should be given to an automobile's year, make, and model, virtually anything is possible. Just go to any custom car show and you'll see Volkswagen Bugs with V-8s, Cadillacs with high-rise manifolds and superchargers, beautifully restored and factory-original Corvettes, high-performance pickup trucks, and a lot more.

Stock Traditions

You may think that maintaining a strictly stock engine compartment would be easy—just clean it up once in awhile and that's it. While this may hold true for newer automobiles with engine paint, stickers, decals, hoses, and belts all in fine condition, what about a 1960s vintage model?

Not that many years ago, yellow ignition wires were the rage. It seemed that every backyard me-

The engine and engine compartment in Jerry McKee's Jaguar are maintained stock. Since he frequently drives the car, restoration efforts are limited to conscientious preventive maintenance. This is not a museum piece nor a national concours contender. The vehicle is used mainly for pleasurable driving; participation in local concours is just an entertaining sideline. The manner in which you maintain, modify, or customize an engine compartment must be weighed against its overall purpose so detailing results achieve both visual and functional goals.

The engine and surrounding compartment space in this 1964 Jaguar E-Type Roadster are maintained in concours condition. It may be driven on rare occasions, but not nearly as much as McKee's. It has been maintained in factory-stock condition in line with concours rules—an option that has paid off in several concours awards.

chanic just *had* to put a set on his or her 1962 Chevy or 1958 Ford F-100. Was it the color that was special, or the quality construction of the wires that made them so popular? In either case, yellow was not in line with traditional ignition wire options.

Attempts to make engine compartments look "cool" are sometimes thwarted by a detailer's lack of imagination. Although one set of yellow ignition wires might not be enough to make an engine compartment look silly, a wacky combination might qualify, such as wires and hoses in exotic but contrasting colors, with mismatched sets of chrome parts and some painted bright silver, oddball-colored valve covers, and a flat black firewall and inner fenders. Which would you prefer, a sparkling clean, stock engine compartment, or an inexpensive poor attempt to cross a pro stock racer, a custom alteration, and a street rod?

Stock is not bad, even for those who never plan to enter a concours competition. Black ignition wires, red heater hoses, an orange- or blue-painted engine, black air cleaner, and flattened gloss black engine compartment can really stand out if all of the parts are cleaned, polished, waxed, and dressed appropriately. When parts break or wear out, you do not have to settle for less than factory-original parts to maintain originality, as new parts or equivalent replicas are available through many new car dealerships and a wide variety of specialty parts businesses that frequently advertise in periodicals like *Hemmings Motor News*.

Getting a neglected or abused engine compartment back to traditional standards may require a bit of legwork. You will have to consult shop manuals and any other available literature for the year, make, and model of your project vehicle. Auto swap meets are generally attended by vendors who specialize in

Race cars, like this dragster, are set up to go fast. They do not have to conform to all of the pollution controls that roadworthy automobiles must comply with. Pro street machines are designed to look a lot like race cars with custom inner fender modifications, powerful engines, and the lack of extra engine compartment accessories. To make your car or truck's engine compartment look like a pro street machine, however, you may have to sacrifice comfort items such as power steering, air conditioning, cruise control, and so on.

An engine sitting in a 1941 Plymouth. It has a few custom touches like a chrome air cleaner, coil clamp, and brake fluid reservoir cap. The effect is rather simple but blends well with a car body that has only had minor custom alterations like frenched headlights, and altered door handles and taillights. Maintaining it in this condition should not require much more than routine cleaning.

such material, and you may even be able to locate an original showroom brochure applicable to your car that shows pictures of what its stock engine compartment is supposed to look like. As you put your car together and need help in deciding which color to paint various parts, check with an autobody paint and supply store jobber, other car club members, professional auto restorers, and even acquaintances who may compete in concours events.

Racing Modifications

Although engine compartments in racing machines might look totally awesome and powerful, looks are secondary to performance. Everything under the hood is there for a reason, and you can bet that if pink polka dot valve covers and leopard skin air cleaners could make cars go faster, every race car would have them.

Heavy-duty is usually the key word for race car competitors. Along with using their sponsors' equipment, they will also use whatever parts perform best for the function required. They do not try to hide wires or hoses for appearance sake, but they may be routed out of the way to gain quick and unencumbered access to spark plugs or adjustment mechanisms. Braided hoses and bolt-on connectors are used because they hold up and perform better than standard hose clamps. The same goes for wiring and other engine parts. The battery is frequently placed in the trunk area, and typically there will be no power steering unit, air conditioning compressor, cruise control device, or any other extra goodies in the engine compartment. Those areas run mighty hot during races and the fewer items in there, the better.

Race car owners, or their mechanics, generally paint engine compartments a bright color, like white. This helps to improve visibility under the hood by reflecting light, and also assists in finding items such as screws, nuts, and so on that accidentally fall during mechanical work. Stainless-steel panels are fre-

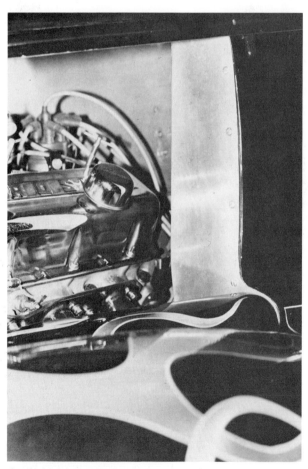

True stainless steel braided hoses and heavy-duty hose connectors are not gracing the front of this race car engine for looks alone. They were installed because of their dependability: they hold up well under high-performance conditions. Racers install engine parts that will improve performance; appearance plays a secondary role.

A polished firewall adorns the engine compartment inside this hot rod. Its clean look blends well with the numerous chrome engine parts. Besides enhancing appearance, another benefit of bright-colored compartment panels is the increased illumination they bring to normally dark and shadowed engine spaces. Many race cars feature engine compartments that are painted white, so mechanics can take advantage of improved light reflection.

Race car engine compartments are not detailed just for appearance sake. All possible engine flaws must be corrected before race time if drivers are expected to finish. Fluid leaks and broken parts are easy to spot on clean and detailed engines. In addition, as mechanics go about cleaning and detailing tasks, they are able to closely inspect engine parts and check them for signs of wear or failure.

Hot rods and street rods are supposed to look much faster than stock machines. Superchargers, fuel injection, dual quads, and three deuces are just some of the features that highlight rodding machines. Along with their go-fast and throaty sounding attributes, bright accessories help to make rods look special, such as the chrome hose end caps, braided wire covers, and chrome water pump on this 1958 Chevrolet motor.

quently installed around engine compartments not only for improved access and visibility to the engine, but also for improved airflow through the engine compartment, increased structural stability, and a decreased weight factor.

So if your race car's engine compartment needs a facelift, consider painting it a light color for improved visibility, and removing every unnecessary item. Since you'll need only the bare essentials, it should be easy to keep them tidy from the start. Use plenty of sturdy hose and wire hold-down brackets to keep things from flopping around while the engine is running, and make sure that you've done everything possible to improve access to the engine and all adjustment mechanisms.

High-performance shops sell heavy-duty items specifically made to keep engines intact under extreme conditions like high-rpm racing. They also sell custom oil pans with larger capacities and a host of

other items designed for high-performance operation. Many parts are chrome or polished alloy because they clean quickly, hold up well, perform as expected, and look good.

Detailing a race car's engine compartment is not done to simply improve its appearance. Meticulous cleaning makes it easier to check the physical condition of parts: Is anything leaking, are cables frayed, are wires cracked, and so on. If an engine is dirty, no mechanic can spot potential problems. All mechanical items must be in top condition if a racer expects to win. Failure to note a problem before a race begins could easily land a car back in the pit after only a lap or two.

Street Rod and Hot Rod Options

Perhaps street rods and hot rods can boast of some of the most visually exciting engine compartments around. They often sport high-performance

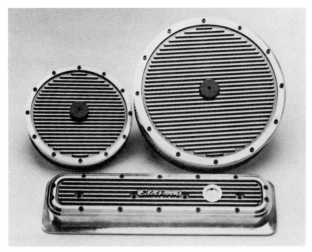

Appearance options frequently selected by rodder enthusiasts include custom air cleaners and valve covers, like these from Edelbrock. Styles range from stock parts painted to match body colors, to those that have been chrome plated, to ribbed designs such as these, to polished billet models. The many choices available allow detailers to select the ones that best suit their personal taste. Photo courtesy Edelbrock Corp.

Superchargers (blowers) are available in numerous sizes to fit a variety of engines. Hot rods and street rods are commonly outfitted with blowers to help them go faster and look better. But adding a supercharger requires more than its installation. The other engine components must be heavy-duty enough to handle the increased power produced by such equipment. Before buying and installing a blower, be sure you understand and can comply with all of the mechanical requirements. Photo courtesy B&M Automotive Products

engine options with splashes of bright colors, lots of chrome or billet, louvered hoods, custom hood scoops, and more. Street rods that are put together right can be eye appealing, fast, street legal, customized, uniquely different, and fun without their owners having to be overly concerned about winning races or maintaining factory originality.

In a manner of speaking, street rods and hot rods may fill a gap between high-performance race cars, like dragsters, and wild custom creations, like the *Batmobile*. Common street rod features include big engines with superchargers, multiple carburetors, headers, chrome or billet pulleys, brackets, and so on. To go along with their custom paint jobs, louvered hoods, and trick wheels, the engine compartments on street rods are kept meticulously clean. Although many of these motors may be dialed in to offer lots of horsepower, the real attraction of most street rod engine compartments is their visual appeal.

Street rod engine compartments can offer an interesting array of options. Some owners like to hide as many engine components as possible to make it look as though the motor just runs on air. Rerouting wires and hoses makes for a tidy and uncluttered engine compartment. Likewise, maintaining a common set of custom wire covers throughout can offer equally appealing results. The main idea to keep in mind when detailing a street rod engine compartment is *uniformity*.

If you install a braided hose from the thermostat housing to the radiator, shouldn't you do the same for the lower hose and heater hoses? If ignition wires are neon pink, shouldn't the spark plug and distributor ends be a matching color? And, would neon pink look good on an orange engine? Wouldn't a blue block or matching neon-colored engine look better? Uniformity is the key. If you start out with a lot of chrome, stay with chrome until the entire engine compartment looks the way you want it.

Look at some prize-winning street rod or hot rod entries at the next car show, or browse through magazines like *Hot Rod* to get a better understanding of what uniformity means. Some of the most stunning rods around have the engines, intake manifolds, valve covers, alternator housings and brackets, radiators, pulleys, and just about everything else painted the same pastel color as the inner fenders, firewall, and car body. The wiring harnesses are routed as far out of the way as possible, hoses are all the braided style, and ignition wires sport a color almost identical to their engines. The total uniformity exhibited by all of these various parts and assemblies inside such engine compartments makes them look outstanding, crisp, and pristine all at the same time.

The manifold should also fit within your chosen theme. Should it be polished to go along with a polished blower and valve covers? Or, should it be painted the same color as those other items? What about the alternator and its bracket? It should match, too, as should all of the other components on the front of your engine.

Street rods, sport trucks, and high-performance muscle cars all rely on their engine compartments to complete their overall look. A well-balanced engine compartment complements the rest of the vehicle and makes the unit look complete.

The firewall, brake master cylinder assembly, and valve cover have been painted the same color as their host automobile. The clean appearance of this engine compartment area is not cluttered with assorted wires or hoses running every which way. They are routed in a specific fashion to create a sense of order.

If choosing complementary colors is difficult for you, refer to a color wheel found at any artist's supply store to see which colors work well together. Compare your findings to the existing color scheme of your car. Then decide which colors will work best for the engine, the ignition wires, the engine accessories, whether or not to use chrome or billet, and so on.

High-performance stores sell a number of goodies for street rodders, and auto-related magazines carry plenty of advertisements for companies that specialize in custom parts. Check out the wide array of cars and trucks featured in these magazines to get an idea of what other enthusiasts have done to their engine compartments. That's what guys like Dan Mycon do. They take some ideas from magazines, others from car shows, add a little of their own imagination, and then set out to detail an engine compartment of their own creation.

Custom Ideas

Street rodders, sport truck enthusiasts, and muscle car buffs customize their vehicles' engine compartments with lots of bright-colored goodies, billet add-ons, and other bolt-on products. True customizers, though, generally rearrange the appearance of their engine compartments by cutting out sections or welding in different ones. They may also elect to install hood scoops, louvers, different grilles, tilt-forward front ends, and so on.

This kind of custom work requires special equipment, some expertise, and a good-sized bank account.

A custom intake manifold from Edelbrock that sports three deuces may be just the thing your engine compartment needs. Manifolds are intricate motor parts that serve a critical function and are also highly visible. High-performance shops sell a variety of intake manifold styles to both improve engine operation and enhance appearance. Photo courtesy Edelbrock Corp.

Rarely does an auto customizer install a factory-original engine in a project car. Hot rods are a good example. What do you suppose Henry Ford would think of some of his 1932 models rolling around today with big, blown V-8s, dual quads, and chrome headers sticking out all over the front? Think he'd mind if a completely different front suspension and steering system were installed, too?

Maverick auto customizers have taken advantage of the technology and equipment available today to combine the best performance features of many newer cars with the unique styles from yesteryear. The high-performance engines of today will not simply bolt on to the frames of cars built forty, fifty, or more years ago. Nor will the standard front ends and steering systems of those older cars function properly with the addition of a new engine without at least some modification.

Once assembled, the engine compartments of most custom-built automobiles resemble those of quality street rods. Blowers and hood scoops hint that underneath the hood or engine cover there may be braided hoses, billet valve covers and accessories, a fresh engine, and lots of shiny and well-maintained engine parts.

Concours d'Elegance

The first time you look at the engine compartment of a Concours d'Elegance winner, you might be surprised at the lack of high gloss and fancy equipment that you would expect to see on a winning street rod or custom show car. That's because concours relishes the art of perfection in rejuvenating and maintaining automobiles to their factory-original condition. George Ridderbusch bought his 1979 Porsche 928 right off of the showroom floor. He began work on it right away, but it wasn't until three years later that he finally won first-place concours honors with it. Dare it be said that at three years old his car was better than new?

In the engine compartment of a concours car, all hoses will exhibit a uniform subtle gloss, each metal part will be painted its factory-original color to the same gloss as when new, wires will be cleaned and routed perfectly as designed, and the radiator fins will be even and symmetrical. This is truly perfection at its best.

Now, not every concours contestant enters a vehicle with these perfect scoring capabilities. In fact, there are classes for those vehicles, like McKee's 1972 XKE, that are driven on a semi-regular basis—on fair-weather summer days. Their engine compartments are allowed to be a little less than perfect, but not much. For more competitive concours classes, like those Ridderbusch enters, entries are most often delivered to events in enclosed trailers and wearing a set of wheels and tires installed just for the trip. Competition tires and wheels are put on after the vehicle is parked in its appropriate location.

The degree of cleanliness found in the engine compartments of Best of Show concours winners is astounding. Not only will you fail to spot a single drop of fluid anywhere, but the chances of finding a particle of dirt are even more remote. To break a tie between two equal winning entries, judges have been known to use cotton swabs to reach deep into engine valleys to find traces of grease, dirt, or grime. The use of white gloves is not unheard of, either.

As much as high gloss, bright colors, and innovative designs are part of street rods and customs,

Beneath a layer of bodyshop dust sits a chrome radiator tank and other engine compartment items. Auto enthusiasts can choose from an endless number of aftermarket parts designed to spruce up almost any engine compartment. The custom chrome fan shroud featured here looks special and helps to make the entire compartment something to admire.

The entire front end has been removed from Dan Mycon's 1948 Chevy and new frame and suspension pieces are being installed. A great deal of work goes into custom modifications and engine detailers must know what they are getting into before tackling such major projects. Frame and suspension work should not be taken lightly. It requires special equipment and expertise to solidly and safely set up, install, and adjust such assemblies.

Thanks to modern technology and the innovative spirit of true auto enthusiasts, the joining of classy body styles from yesteryear with the powerful and reliable powerplants, transmissions, suspensions, and brakes of today has been made possible. Kits are also available for the installation of big V-8 engines in older cars, as well as brackets and supports for securing larger radiators which are needed to cool high-performance motors.

subtle perfection is the mainstay of the concours automobile. Total originality is mandatory, and points are deducted for those parts not offered as factory-direct options at the time a particular vehicle was sold. Therefore, when detailing a concours engine compartment, great care must be taken to preserve factory markings. These can include blotches of paint intentionally placed on parts, bolts, and other fasteners by factory supervisors as a means to acknowledge that the item has been properly installed and torqued to specifications. Or, they could be actual stenciled, painted-on letters or numbers. During an in-depth detail or restoration, photos should be taken of such markings so they can be duplicated if removed for bodywork, or if they are damaged or inadvertently erased.

Changing the design of a concours contender's engine compartment is taboo. The only acceptable changes, for the most part, are limited to aligning hose clamps so they all adjust in the same direction and at the same angle, lining up slots on screws so they all point in the same direction, and removing factory-installed Cosmolene protectant from inner fender aprons and other parts.

The engine compartments on concours cars need a lot of attention, especially just prior to an event. Ridderbusch and McKee like the job Meguiar's #7 does in polishing painted surfaces on the underside of the hood, fender aprons, and firewall. Wax is seldom, if ever, used on those areas, they say, because it could tend to build up and dull after awhile.

Meguiar's #7 brings out just the right amount of gloss needed and keeps paint looking that way for about three days, long enough to survive judging.

If factory-original perfection is what you're after, then Concours d'Elegance may be an avenue to pursue. You could become a perfectionist, like Ridderbusch, or enter the classes for cars frequently driven, like McKee. Either way, you are bound to meet other enthusiasts with an interest in the same type and style automobile as yours. Because of your shared interest, there might be a wealth of information and knowledge waiting in the wings to be passed around and enjoyed.

Overview

Except for those vehicles that must remain in original condition, appearance options are generally wide open. You can follow suit with the rest of the crowd, or break away and do your own thing. The first engine detailer to paint the block, heads, valve covers, and intake manifold the same color as the body on his car must have initially received some wild looks from other enthusiasts. But, the fact remains that a trend

The custom firewall on Mycon's Chevy blends well with the newly installed bigger engine. The sheet metal piece was fabricated just for this installation. If you intend to perform custom sheet metal work on the engine compartment of your car or truck, be sure to practice first on old scrap doors, hoods, deck lids, and the like.

Concours d'Elegance is where detailers can show off their very best work; it is the ultimate test of detail skills. The engine compartment in John Hall's Jaguar is perfect. Notice how the hose clamps are lined up perfectly, hoses and wires are tidy, brightwork is polished to a balanced gloss, and there is not a speck of dirt or oil anywhere.

was started and is now followed by many avid engine compartment detailers and enthusiasts.

Nobody knows what the next trend will be. Along with body-colored engines, billet accessories are popular as well. Years ago, chrome was the rage. Next, it may be monochromatic engine compartments where one predominant color will cover everything, including belts, wires, cables, you name it.

Selecting the right look for your car may require some effort. Once your engine compartment has been cleaned to perfection, study its layout to see what possibilities may exist. Look at lots of car and truck magazines for ideas, and attend a few car shows. Talk to other auto enthusiasts, and window shop at the local high-performance store. If possible, visit a professional auto restorer and ask if you can observe work being done to get an idea of what you may be up against, especially when a neglected car or truck is involved.

Go to swap meets and pick up copies of old shop manuals and literature that pertain to your make and model of vehicle. If you want to maintain originality, these materials will be valuable to you. Even if you plan to do a bit of custom work, the information provided in some shop manuals may be beneficial when you gear up to cut away panels or alter suspension or steering mechanisms.

By all means, have some idea of what it is you want to accomplish before jumping in with both feet. Old garages and barns are filled with basket-case cars and trucks that were someday going to be restored. Those enthusiasts may have started their project without a road map or even a clue as to how or what they wanted to accomplish, or any idea of the amount of work involved. Don't fall into the same trap with your engine compartment project. Put together a game plan at the outset, follow it, alter it along the way as necessary, and then finally conclude the project feeling that you have accomplished what you set out to achieve.

Chapter 8

Parts Considerations

To make an engine or engine compartment look different than an original factory offering, alterations must be made. This might be accomplished by painting original parts a different color, having them chromed, or by replacing them with similar parts that will maintain or improve performance but are more visually appealing.

Automobile enthusiasts have enjoyed this sort of detailing for years. Some of their changes are subtle, like the installation of chrome air cleaner housings and oil filler caps. Others are more extreme, like the changeover to high-performance parts including superchargers, headers, braided hoses and wires, billet accessories, and so on.

This engine uses assorted aftermarket engine parts available at many high-performance shops. Billet pulleys, air conditioning compressor, alternator, and other components give it a pleasing appearance. All an enthusiast has to do is shop around at high-performance stores and look in auto magazines to get an idea of all the custom engine accessories available.

Chrome is the theme expressed in this engine compartment. All parts and accessories are outfitted to coordinate with a basic brightwork appearance. Braided wires and hoses complement chrome parts to help make everything look balanced, like it all belongs together. It is like a three-piece suit combination where shoes, tie, and socks must be coordinated with the coat, jacket, and vest. White socks and running shoes would no more blend with a three-piece pinstripe suit than would a set of stock orange valve covers on this engine.

Auto parts stores commonly offer aftermarket add-on accessories for a wide range of applications. For a while, the rage was chrome and enthusiasts could easily find such items as custom chrome air cleaners, valve covers, spark plug wire brackets, and the like. Billet parts are popular these days, with high-performance street rods and hot rods sporting everything from billet pulleys to thermostat housings.

The availability of custom aftermarket add-on products has never been better. A host of high-performance shops and mail-order companies offer just about any part you would ever want. In addition, plenty of local and regional companies offer services that will transform stock parts into colorful or brightly polished ones in just a few days.

For those involved in restoration projects, original parts may be of utmost concern. Fortunately, the increased interest in older automobiles over the last few years has led to a relative boon of new companies that specialize in selling new, used, rejuvenated, and quality reproduction parts for classic and vintage cars and trucks. Many such businesses frequently advertise in *Hemmings* and other periodicals geared toward auto restoration.

A number of specialized wrecking yards (auto dismantlers) have also sprung up. They deal in older cars and trucks and pride themselves on keeping tidy yards where customers are only allowed to browse. All parts are dismantled by employees to ensure that other salvageable parts are not inadvertently damaged by eager patrons. Although used parts may cost less than reproductions, you should factor in rejuvenation costs that may include sandblasting or bead blasting, polishing, and mechanical repair of internal components.

Chrome

Because of increasingly stringent regulations governing the use and disposal of hazardous materials, the cost of chrome plating has gone up. Small parts, like an alternator bracket or pulley, may cost

Common aftermarket engine compartment parts include air cleaners, valve covers, oil filler caps, breathers, brackets, and so on. Engine detailers can start with items like these and gradually add new parts as their budget allows. Be sure to stay with one motif and purchase only the items you really want.

This engine is being put together and has yet to be detailed or polished. Dan Mycon has chosen to go with billet accessories. Along with the billet pulleys, alternator, and bracket, he has installed a custom billet timing marker and used Allen head screws and bolts for the installation.

105

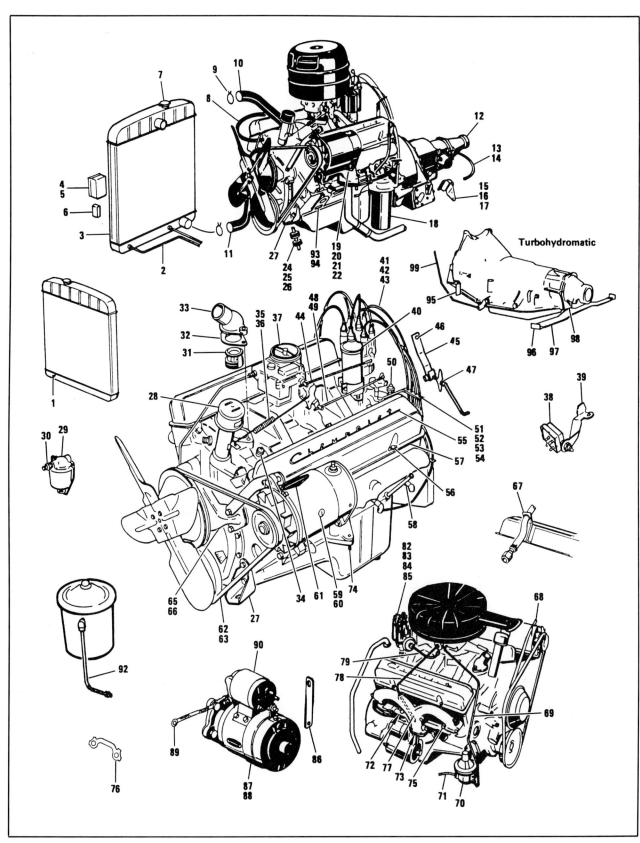

Turbohydromatic

For enthusiasts restoring stock engine compartments, a number of companies make stock part reproductions and exact replicas. Some are even able to offer NOS (new-old-stock) parts. This is a sample of available engine parts for 1955-57 Chevrolets through Drake Restoration Supplies. Reprinted courtesy Drake Restoration Supplies

between $50 and $100 to plate, while larger items, like bumpers, may run as much as $400 for show car perfection.

Over-the-counter chrome parts sold on the retail market may cost less than custom-plating work because those parts were plated on an assembly line at a production plant. It is cheaper to plate 100 identical parts than it is to process individual items. Therefore, before enlisting the services of a local chrome-plating company, check around with different high-performance shops to see if they carry similar parts at lower prices. Be sure the quality of those parts meets your standards, and that the design and bolt holes line up exactly the same as your stock parts.

Just about any metal item in an engine compartment can be chromed, from radiator tanks and headers to valve covers and brake fluid reservoirs. There are even companies like Gary's Plastic Chrome Plating, Incorporated, of Westland, Michigan, that specialize in chrome-like plating for plastic parts. These companies advertise in *Hemmings* and a number of other auto-related magazines. Since each plastic part is individually plated and returned to you, as opposed to an exchange program, plastic chrome-plating companies recommend you send them your best parts, those free of deep scratches, cracks, or other blemishes. You may be able to locate chrome engine and engine compartment parts through classified ads in newspapers, weekly shoppers, and monthly automotive periodicals. Auto swap meets are generally attended by enthusiasts who have

Chrome has been a custom automotive mainstay for years. Its brilliant shine is very appealing and easy to maintain. Virtually any metal part can be chrome plated, including hood springs and supports. The cost involved has increased in the last few years, however, so be sure to get estimates before transporting a load of parts to the local plating factory.

Chrome oil pans are commonly available through high-performance shops and some auto parts stores. They are not difficult to install, although removing an old pan might present some oil drip problems. Be sure to completely clean old gasket material off of the block before installing a new gasket and oil pan assembly.

This custom chrome heater hose bracket blends well with its surrounding chrome engine compartment pieces. To continue a balanced effect, other hose and wire brackets should also be chrome.

107

Mycon installed chrome pieces on the nose of his 1948 Chevy to brighten up the engine compartment's sheet metal work. Every day is like Christmas to an auto enthusiast when shiny new parts arrive, ready to be installed on a favorite car or truck.

cleaned out their garages or shops and are willing to sell parts they no longer need. Parts in solid but cosmetically neglected shape might easily be brought back to excellent condition by meticulous polishing with Happich Simichrome or #0000 steel wool.

Polished Aluminum and Billet

Highly polished aluminum and alloy billet parts are not quite as mirror-like as chrome parts, but do give a clean, bright appearance to tidy engine compartments. High-performance shops and a host of mail-order parts dealers offer wide assortments of billet engine accessories that include, but are not limited to, alternators, air conditioning compressors, brackets, pulleys, valve covers, air cleaners, dipsticks, and thermostat housings.

B&M Automotive Products of Chatsworth, California, offers many trick items in highly polished finishes. In their catalog, you can find polished superchargers along with their respective chrome and polished pulleys and brackets. In addition, they offer polished-aluminum swivel thermostat housings, billet alternator fans, polished breathers, gold-chromate-finished harmonic balancers, a blemish-free diecast-aluminum blower-style air filter scoop, and assorted custom valve covers.

Spectre Industries of San Jose, California, adds color to their polished parts by way of anodized

Just one of many billet thermostat housing styles available through high-performance shops and mail order companies. Various engine designs require different thermostat housing shapes and sizes. Some bend at angles, and others swivel into position to join the radiator hose end.

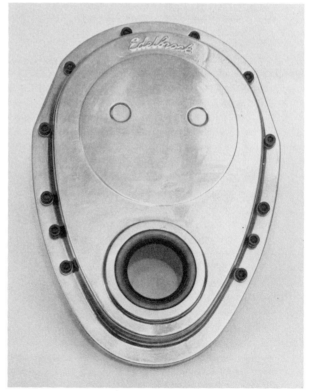

A sample of aftermarket parts available through Edelbrock includes this polished timing chain cover. Photo courtesy Edelbrock Corp.

	PART NO.	SIZE	APPLICATION
STRAIGHT HOSE END	HE-000-04	# 4	FUEL LINES, FUEL AND OIL PRESSURE GAUGES
	HE-000-06	# 6	FUEL LINES
	HE-000-08	# 8	FUEL, REMOTE OIL FILTER LINES
	HE-000-10	#10	DRY SUMP, REMOTE OIL FILTER LINES
	HE-000-12	#12	DRY SUMP, REMOTE OIL FILTER LINES
	HE-000-16	#16	DRY SUMP AND WATER LINES
	HE-000-20	#20	WATER LINES
45° HOSE END	HE-045-04	# 4	FUEL LINES, FUEL AND OIL PRESSURE GAUGES
	HE-045-06	# 6	FUEL LINES
	HE-045-08	# 8	FUEL, REMOTE OIL FILTER LINES
	HE-045-10	#10	DRY SUMP, REMOTE OIL FILTER LINES
	HE-045-12	#12	DRY SUMP, REMOTE OIL FILTER LINES
	HE-045-16	#16	DRY SUMP AND WATER LINES
	HE-045-20	#20	WATER LINES
90° HOSE END	HE-090-04	# 4	FUEL LINES, FUEL AND OIL PRESSURE GAUGES
	HE-090-06	# 6	FUEL LINES
	HE-090-08	# 8	FUEL, REMOTE OIL FILTER LINES
	HE-090-10	#10	DRY SUMP, REMOTE OIL FILTER LINES
	HE-090-12	#12	DRY SUMP, REMOTE OIL FILTER LINES
	HE-090-16	#16	DRY SUMP AND WATER LINES
	HE-090-20	#20	WATER LINES
120° HOSE END	HE-120-06	# 6	FUEL LINES
	HE-120-08	# 8	FUEL, REMOTE OIL FILTER LINES
	HE-120-10	#10	DRY SUMP, REMOTE OIL FILTER LINES
	HE-120-12	#12	DRY SUMP, REMOTE OIL FILTER LINES
	HE-120-16	#16	DRY SUMP AND WATER LINES
	HE-120-20	#20	WATER LINES
180° HOSE END	HE-180-06	# 6	FUEL LINES
	HE-180-08	# 8	FUEL, REMOTE OIL FILTER LINES
	HE-180-10	#10	DRY SUMP, REMOTE OIL FILTER LINES
	HE-180-12	#12	DRY SUMP, REMOTE OIL FILTER LINES
	HE-180-16	#16	DRY SUMP AND WATER LINES
	HE-180-20	#20	WATER LINES

Stainless steel braided hoses are popular additions to many hot rod, street rod, muscle car, and pro street engines. Along with the hoses, special hose ends may be needed in order to make proper line connections. These are a few of the hose ends offered by Keith Black Systems. Photo courtesy Keith Black Systems

accessories. Along with polished custom air cleaners, they offer anodized models in red, gold, and blue. Snap-on caps are available in the same colors for components such as power steering units, vacuum advances, radiators, breathers, oil fillers, and bolt heads. Their catalog also carries custom-anodized dipsticks in red, gold, blue, chrome, or aluminum for small-block Chevy engines.

The Edelbrock Corp. manufactures intake manifolds and "Total Power Packages" for many different engines. A typical package might include a carburetor, intake manifold, and camshaft, with some requiring a special tubular exhaust system. Edelbrock also offers a variety of racing products like fuel pumps, fuel lines and accessories, aluminum water pumps, and cylinder heads. Aluminum valve covers, breathers, hold-downs, and air cleaners are also available.

The best way to learn about all of the polished-aluminum and billet engine accessories available for your engine is to read through display and classified advertisements in car and truck magazines. These ads will tell you how to acquire catalogs, either through toll-free telephone numbers or mailing addresses. Some catalogs are quite thick and very detailed, and frequently include lots of color photographs of the parts inventory. Because of printing costs, some catalogs are no longer free. Their costs range from about $1 to as much as $5 for the more descriptive materials.

Hoses and Wires

Braided hoses, like those you would expect to see on a jet airplane's engine assembly, have become a popular high-performance engine plumbing alternative. Keith Black Systems of South Gate, California, offers stainless steel braided hose for just about any engine application. Different sizes are available for use as oil lines, fuel lines, pressure gauges, vacuum lines, and heater and radiator hoses. Hose end assemblies incorporate the aviation nipple-and-cutter

Braided hoses and their colorful red and blue hose ends are compatible with this big supercharger. More than one style of hose end coupling is available, however, so you'll need to read about each style before choosing one. Real braided lines, as opposed to covers, are designed as heavy-duty engine add-ons. If that is what you are really after, then learn as much as possible about the brand and design to be certain it meets your expectations.

An assortment of Heli-Tube spirally cut cable and wire wrap covers. They come in a number of different colors and sizes and can be used to insulate spark plug wires, prevent chafing on brake lines, and as a bundling system and chafe guard for other engine compartment wiring needs. Photo courtesy M. M. Newman Corp.

method to prevent blowoff and leakage problems. They come in straight and angled designs from 45 degrees to 180 degrees. To round out their assortment of braided hoses and hose ends, Keith Black Systems carries a full line of adapter fittings, from flare plugs and tube nuts to tees and elbows.

Since not everyone may be able to completely outfit their engines with braided hoses and connections, clamp cover kits and stainless overbraid coverings are available from a number of aftermarket sources, including Keith Black Systems. At first glance, clamp covers look just like hose ends with screw-on connectors shaped like bolts. However, clamp covers are hollow and feature an opening on one side which is slipped over regular hose clamps. Unless a curious onlooker checks under the clamp cover, he or she may not know the connector was an imitation. Overbraid coverings are simply braided covers into which hoses or wires are inserted.

A variety of covers are made to go over wires. Some come in colors like yellow, red, blue, silver, and black. Many are split spiral in design to allow branch wires to exit bundles or harnesses. Others are solid with convoluted ribs to offer plenty of chafing protection. Some covers are even designed to go over existing hoses.

Spectre Industries offers a full line of quality hose covers and connectors for vacuum lines, wind-

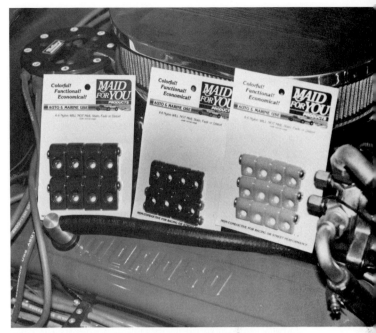

Line clamps are available in several colors and are used to keep wires securely positioned along fender aprons, firewalls, and other engine compartment surfaces. Various sizes will accommodate brake lines, rubber or braided hoses, and assorted wire widths. Clamps courtesy The Eastwood Company

Spectre Industries carries aftermarket engine compartment parts for detailers who cannot afford genuine stainless-steel braided lines and connections. This is a sample of their braided line covers and connectors.

Although some imitation connections simply snap over existing hose clamps, these fittings are actually secured in place. Photo courtesy Spectre Industries

shield wiper fluid lines, spark plug wires, heater and radiator hoses, fuel lines, power steering hoses, and air conditioning lines. For those who want their engine compartment to sport the appearance of braided hoses and colored connectors without the expense involved, this may be a viable alternative.

Taylor Cable Products, Incorporated, of Grandview, Missouri, specializes in automotive ignition and other wiring needs. They offer a full line of spark plug wires in such colors as black, red, yellow, blue, hot orange, hot lime, and hot pink. They also carry wire separators in the same range of colors. Along with a wide assortment of wire connectors and other electrical items, Taylor makes Diamondback Shielded Battery Cables which are designed to complement engines outfitted with braided-line engine accessories.

Other Goodies

A number of companies, like Earl's Performance Products of Long Beach, California, offer plenty of small detail items that can make your billet and braided hose-equipped engine compartment look finished. These would include billet aluminum dipstick handles that are tightened onto existing units with a set screw. Other parts include intake manifold,

water pump, thermostat housing, and timing cover hardware with Allen head bolts and colored cup washers. You can also get billet aluminum carburetor, distributor, and valve cover hold-downs, as well as billet V-8 carburetor return spring brackets and low-pressure aluminum tubing in red, blue, or non-anodized colors.

If you do enough shopping around at various high-performance centers and read plenty of auto-related magazines, you'll soon notice that there are numerous sources that offer just about any engine compartment accessory imaginable. Don't forget new car dealership parts departments. Major auto manufacturers acknowledge that many customers are interested in high-performance and appearance options and are now offering those kinds of parts and accessories to their dealers.

Ford Motorsport Performance Equipment, for example, is now available through Ford dealerships. Along with a chrome fan and air cleaner, and assorted valve covers, Ford makes such factory high-performance products as manifolds, cams and lifters, heads, crankshafts, headers, and high-energy ignition sys-

Spark plug wires no longer have to be just plain black. Taylor Cable Products offers an array of spark plug wire colors with unique end styles. They also offer a wide selection of ties, electrical wire, terminals, boots, and lots of other automotive electrical parts. Wires courtesy Taylor Cable Products

Converting to a new V-8 engine may require special brackets to accommodate alternators and other accessories. Check with local high-performance shops to see if the brackets you need are available in billet or chrome. Some bracket assemblies are made especially for specific engine conversions.

tems. They also advertise chrome engine oil and transmission pans.

If you install a blower on top of your car's engine, chances are you'll need to cut a hole in the hood and install a hood scoop or purchase a new hood complete with a built-in scoop. Hardwood Industries, Incorporated, of Tyler, Texas, manufactures aerodynamically designed hood scoops for just about every application. They also produce a number of fiberglass bolt-on hoods and a selection of Chevrolet Camaro and Beretta front ends.

If you have opted for a high-performance, street rod, hot rod, or other custom theme, you may have to consider installing an electric fan. This is the only type of fan that will work on some custom modifications. Scotts Manufacturing Company of Valencia, California, offers a number of different electric fan and motor sizes for all kinds of high-performance and custom needs, including models for 4x4s, motorhomes, and off-road vehicles.

Headers are another type of engine alteration that many auto enthusiasts insist upon for their vehicle's engine compartment. Besides helping engines to perform better, headers make them sound good. Although there is some concern about the increased heat created by blown engines with headers, you must also realize that heat needs to stay in the exhaust system to help gases maintain high velocity. To assist in keeping exhaust heat under control, many race car drivers, owners, and mechanics have installed Thermo-Tec's (of Berea, Ohio)

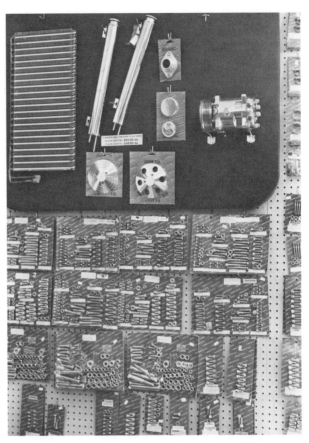

To make the installation of new parts uniform and complete, use bolts, screws, and other fasteners that match the new parts. In other words, use chrome bolts with chrome parts and polished bolts with billet or polished alloy components. These assorted nuts, bolts, and washers are packaged for use with specific installations.

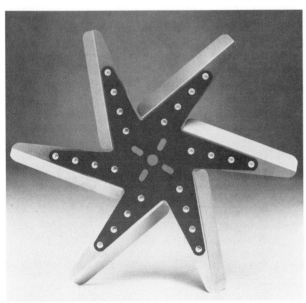

Custom fans not only help to improve engine cooling, but some designs look nice, too. This is a 1000 Series Fan from Flex-a-lite Consolidated. It is designed for high-performance driving on engines without air conditioning. Fans are available with rigid aluminum, chrome, or stainless-steel blades. Photo courtesy Flex-a-lite Consolidated

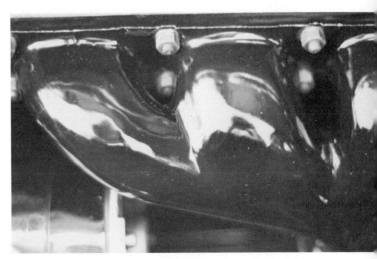

The stock porcelain exhaust manifold on John Hall's 1959 Jaguar. Finding an original part like this for a restoration project may be difficult. If street rod enthusiasts feel like it is Christmas when they buy new goodies for their ride, imagine how a serious restorer would feel finding a rare stock porcelain exhaust manifold or other rare part.

113

Locating parts for restoration projects that are not necessarily slated for national concours will be a lot easier than hunting for factory-original ones. Publications like Hemmings *are filled with advertisements from companies such as Year One, Inc. and Metro Moulded Parts, Incorporated, that specialize in reproduction automobile parts. Virtually any part you need to put a vehicle back together should be available through a reproduction auto parts company.*

Exhaust Insulating wrap around headers. Thermo-Tec also makes blankets for stock exhaust manifolds.

When Art Wentworth looks back on all of the cars he has owned, from twenty-six brand-new Corvettes, to Jaguars and Rolls-Royce's, to a nifty 1970 Buick GS 455 convertible, he can remember what it was like to celebrate Christmas in the middle of summer. How? By purchasing new goodies for his cars. Although he prefers engine compartments that are clean and natural, every once in a while he has a compulsion to buy new decals for an air cleaner or a complete set of heater and radiator hoses.

Other enthusiasts, like Dan Mycon, would never be happy with just a set of decals. He would (and does) want every trick new item that comes down the pike. Billet is in, so Mycon's project car will sport a lot of it. To him, every day is Christmas when a package arrives from one of his favorite high-performance outlets like RB's Obsolete Automotive, Incorporated.

It's fun to set up an engine compartment to look the way you want it to. Sure, it costs a few dollars, but look at the bang you get out of your buck! Be as frugal as you have to, by shopping at swap meets and checking classified ads, but more important, have a vision of what you want your engine compartment to look like when you're done. Buying something that is only *close* to what you really want is neither frugal nor effective. Wait until you can afford exactly what you want and then install it with pride and the knowledge that it will stay in place for a long time. (See Appendices for source details.)

Locating Original Parts

For engine compartment detailers putting together an original machine, all of the billet goodies and neon-colored parts are essentially a waste of time. Classic car restorers want parts that look, feel, and smell like factory-originals. To them, Christmas is finding a battery tray or inner fender assembly that hasn't totally rusted through or disintegrated.

Finding usable antique and classic parts is not always easy, though, especially those that had been stamped at the factory with specific identification numbers on their original housings. Any knowledgeable vintage car aficionado or concours judge would be quick to point out ID number discrepancies. But original parts can be found for those restorers who are determined, patient, and thorough. Check with local vintage auto dismantlers to see if they have what you need. Be sure to study any available shop manuals or other material that can clearly show you what specific parts look like and how they can be accurately identified.

After scouring vintage auto dismantling facilities, talk to members of a car club that specializes in the same year, make, and model automobile as yours. Who better than a group of avid enthusiasts to know where parts might be hiding, or who may have some extras sitting in their garage? Next, try auto swap meets. Talk to venders who seem to have interesting assortments of items for sale. Maybe they have a cache of old cars sitting in a pasture, barn, or backyard that still sport original parts in decent condition.

Local sources may include professional auto restorers with contacts throughout the country, or an avid car enthusiast who is owner and operator of a local body and paint shop. Most of the people involved in auto-related businesses really enjoy automobiles and have interests that stretch beyond their skills to earn a living. Their contacts might know where you could find those needed parts as well.

Auto-related magazines and periodicals are starting to carry more and more classified advertisements from subscribers. As word gets around that vintage car parts are becoming scarce and therefore more valuable, those who had more or less forgotten about baskets of old car parts in their sheds, garages, and shops are beginning to sift through and sell what is no longer of value to them.

Once located and purchased, parts must be evaluated for overall appearance and usability. The housing on a generator, for example, might appear to be in fine shape, but what about the windings and other internal mechanisms? If a case is good enough to restore, have a professional rebuild the guts so that the unit will function as expected and still look original. Consult with an autobody paint and supply store jobber to determine correct paint colors for use on outer housings to maintain originality.

George Ridderbusch discovered a problem in locating parts for his 1957 Porsche 356 A project car. He took advantage of contacts he had in the Porsche Club of America to locate what he needed. This kind

MOTOR MOUNTS
1932-1970

Motor mounts should always be replaced in pairs or sets along with transmission mounts. Age, oil and grease cause deterioration of the rubber leading to vibration, chatter and failure. No restoration is complete without new motor mounts.

The listing below is a general guide only. Popular mounts are available at listed prices. Some mounts are available at higher prices and others are not available.

In some cases, we may ask you to send your old one in as a sample.

FRONT OR SIDE MOUNT			REAR OR TRANS MOUNT		
year	price each	qty.	year	price each	qty.
Buick					
40-47	$50 exchange	2	41-47	$75 exchange	2
48-60	$35	2	48-59	$30	2
61-70	$35	2	64-70	$25	1
(except '64 LeSabre and '66 Wildcat & Electra)					
Buick Special					
61-63 V8	$25	2	64-70	$25	1
65-67 V8					
300, 340	$35	2			
Cadillac					
37-48 V8	$25	2	39-48	$30	1
38-40 V16	$25	2	—	—	—
49	$50	2	49	$30	1
50-70	$40	2	50-60	$30	1
Chevrolet Full Size					
35-51 Front	$30	2	37-51	$35	1
35-51 Side	$15	2	—	—	—
(Specify if side mount is "W" shaped or rectangular)					
52-54	$20	4	52-54	$25	1
55-57	$20 for set of 8		55-57	$25	2
58-69	$20	2	58-69	$20	1
Chrysler					
37-54 6 cyl	$30	1	41-54 6 cyl	$40 for set of 4	
51-53 8 cyl	$40 for set of 4		51-52 8 cyl	$40 for set of 4	
			53 8 cyl	$30 for set of 4	
54-56 8 cyl	not available		55-56 8 cyl	$30	1
57-59	$35	2	57-59	$25	1
60-64	not available		60-64	$25	1
Corvette					
53-62	$20	4	53-62	$25	1
Dodge Desoto/Plymouth					
35-56 There are 2 basic types of front and rear motor mounts used.					
1. Bar Type - Approximately 1" x 6", $30 each, 1 used per car					
2. Round Cushion Type - Used in sets of 4 at front or rear, $40 set					
Note: Dodge 56 V8, Front mounts not available					
Desoto 55-56, Front mounts not available					
Plymouth 56 V8, Front mounts not available					
57-70	$25	2	57-70	$25	1

FRONT OR SIDE MOUNT			REAR OR TRANS MOUNT		
year	price each	qty.	year	price each	qty.
Ford					
32-41	$25 for set of 4		37-41	$25 for set of 4	
42-48 V8	$25 for set of 4		42-48 V8	$35	1
49-68	$25	2	49-68	$25	1
Ford Fairlane					
65-69	$25	2	62-69	$20	1
Hudson					
48-54	$25	2	—	—	—
Lincoln					
49-51	$35 for set of 4		49	$75 exchange	1
52-58	$20 lower	2	—	—	—
Mercury					
39-41	$25 for set of 4		39-41	$25 for set of 4	
42-48	$25 for set of 4		42-48	$35	1
49-51	$35 for set of 4		50-51	$25*	1
52-70	$25	2	52-70	$25	1
Olds					
37-53 (upper)	$30	1	49-53	$30	2
37-53 (lower)	$30	1			
54-64	$30	1	54-64	$30	2
65-69	$25	2	65-69	$25	2
Olds F-85 Cutlass					
61-63 V8	$25	2			
64-69 V8					
(exc. 400")	$35	2	64-69	$25	1
Packard					
Six & Small 8					
35-50	$20	1	38-50	$45	2
356" 8 40-50	$20	1	40-50	$45	2
Pontiac					
40-58	$30	1	40-58	$30	2
59-64	$30	2	59-64	$25	1
64-70	$25	2	64-70	$25	1
Tempest					
8 cyl. 63-69	$25	2	64-70	$25	1

49-50 standard transmission $75 exchange

Add 10% for postage and handling in continental U.S.
($3.00 minimum, $30.00 maximum)

To Order: State make, year, model, body style, engine type and cubic inch, transmission type, left, right, side, rear, or transmission.
Inquire for Studebaker, Nash, AMC, Kaiser-Frazer, truck and other mounts not listed.
See Full Listing of Engine Parts on pages 33-39.

Parts for cars and trucks from the 1920s and 1930s are available, as well as for newer vehicles. This page from Kanter Auto Products' Catalog shows part of their motor mount inventory. Do your homework before starting to tear down an old project vehicle. Be certain you are able to find all of the needed parts and that the cost is within your budget. Courtesy Kanter Auto Products

HOOD BUMPERS

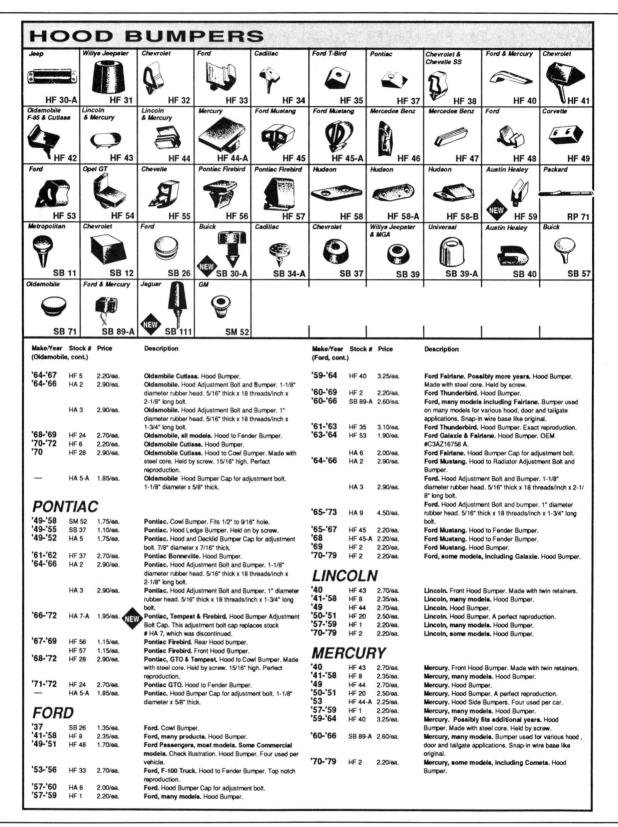

Make/Year (Oldsmobile, cont.)	Stock #	Price	Description
'64-'67	HF 5	2.20/ea.	Oldsmbile Cutlass. Hood Bumper.
'64-'66	HA 2	2.90/ea.	Oldsmobile. Hood Adjustment Bolt and Bumper. 1-1/8" diameter rubber head. 5/16" thick x 18 threads/inch x 2-1/8" long bolt.
	HA 3	2.90/ea.	Oldsmobile. Hood Adjustment Bolt and Bumper. 1" diameter rubber head. 5/16" thick x 18 threads/inch x 1-3/4" long bolt.
'68-'69	HF 24	2.70/ea.	Oldsmobile, all models. Hood to Fender Bumper.
'70-'72	HF 6	2.20/ea.	Oldsmobile Cutlass. Hood Bumper.
'70	HF 28	2.90/ea.	Oldsmobile Cutlass. Hood to Cowl Bumper. Made with steel core. Held by screw. 15/16" high. Perfect reproduction.
—	HA 5-A	1.85/ea.	Oldsmobile. Hood Bumper Cap for adjustment bolt. 1-1/8" diameter x 5/8" thick.

PONTIAC

Make/Year	Stock #	Price	Description
'49-'58	SM 52	1.75/ea.	Pontiac. Cowl Bumper. Fits 1/2" to 9/16" hole.
'49-'55	SB 37	1.10/ea.	Pontiac. Hood Ledge Bumper. Held on by screw.
'49-'52	HA 5	1.75/ea.	Pontiac. Hood and Decklid Bumper Cap for adjustment bolt. 7/8" diameter x 7/16" thick.
'61-'62	HF 37	2.70/ea.	Pontiac Bonneville. Hood Bumper.
'64-'66	HA 2	2.90/ea.	Pontiac. Hood Adjustment Bolt and Bumper. 1-1/8" diameter rubber head. 5/16" thick x 18 threads/inch x 2-1/8" long bolt.
	HA 3	2.90/ea.	Pontiac. Hood Adjustment Bolt and Bumper. 1" diameter rubber head. 5/16" thick x 18 threads/inch x 1-3/4" long bolt.
'66-'72	HA 7-A	1.95/ea.	Pontiac, Tempest & Firebird. Hood Bumper Adjustment Bolt Cap. This adjustment bolt cap replaces stock # HA 7, which was discontinued.
'67-'69	HF 56	1.15/ea.	Pontiac Firebird. Rear Hood bumper.
	HF 57	1.15/ea.	Pontiac Firebird. Front Hood Bumper.
'68-'72	HF 28	2.90/ea.	Pontiac, GTO & Tempest. Hood to Cowl Bumper. Made with steel core. Held by screw. 15/16" high. Perfect reproduction.
'71-'72	HF 24	2.70/ea.	Pontiac GTO. Hood to Fender Bumper.
—	HA 5-A	1.85/ea.	Pontiac. Hood Bumper Cap for adjustment bolt. 1-1/8" diameter x 5/8" thick.

FORD

Make/Year	Stock #	Price	Description
'37	SB 26	1.35/ea.	Ford. Cowl Bumper.
'41-'58	HF 8	2.35/ea.	Ford, many products. Hood Bumper.
'49-'51	HF 48	1.70/ea.	Ford Passengers, most models. Some Commercial models. Check illustration. Hood Bumper. Four used per vehicle.
'53-'56	HF 33	2.70/ea.	Ford, F-100 Truck. Hood to Fender Bumper. Top notch reproduction.
'57-'60	HA 6	2.00/ea.	Ford. Hood Bumper Cap for adjustment bolt.
'57-'59	HF 1	2.20/ea.	Ford, many models. Hood Bumper.

Make/Year (Ford, cont.)	Stock #	Price	Description
'59-'64	HF 40	3.25/ea.	Ford Fairlane. Possibly more years. Hood Bumper. Made with steel core. Held by screw.
'60-'69	HF 2	2.20/ea.	Ford Thunderbird. Hood Bumper.
'60-'66	SB 89-A	2.60/ea.	Ford, many models including Fairlane. Bumper used on many models for various hood, door and tailgate applications. Snap-in wire base like original.
'61-'63	HF 35	3.10/ea.	Ford Thunderbird. Hood Bumper. Exact reproduction.
'63-'64	HF 53	1.90/ea.	Ford Galaxie & Fairlane. Hood Bumper. OEM #C3AZ16758 A.
	HA 6	2.00/ea.	Ford Fairlane. Hood Bumper Cap for adjustment bolt.
'64-'66	HA 2	2.90/ea.	Ford Mustang. Hood to Radiator Adjustment Bolt and Bumper.
	HA 3	2.90/ea.	Ford. Hood Adjustment Bolt and Bumper. 1-1/8" diameter rubber head. 5/16" thick x 18 threads/inch x 2-1/8" long bolt.
'65-'73	HA 9	4.50/ea.	Ford. Hood Adjustment Bolt and bumper. 1" diameter rubber head. 5/16" thick x 18 threads/inch x 1-3/4" long bolt.
'65-'67	HF 45	2.20/ea.	Ford Mustang. Hood to Fender Bumper.
'68	HF 45-A	2.20/ea.	Ford Mustang. Hood to Fender Bumper.
'69	HF 2	2.20/ea.	Ford Mustang. Hood Bumper.
'70-'79	HF 2	2.20/ea.	Ford, some models, including Galaxie. Hood Bumper.

LINCOLN

Make/Year	Stock #	Price	Description
'40	HF 43	2.70/ea.	Lincoln. Front Hood Bumper. Made with twin retainers.
'41-'58	HF 8	2.35/ea.	Lincoln, many models. Hood Bumper.
'49	HF 44	2.70/ea.	Lincoln. Hood Bumper.
'50-'51	HF 20	2.50/ea.	Lincoln. Hood Bumper. A perfect reproduction.
'57-'59	HF 1	2.20/ea.	Lincoln, many models. Hood Bumper.
'70-'79	HF 2	2.20/ea.	Lincoln, some models. Hood Bumper.

MERCURY

Make/Year	Stock #	Price	Description
'40	HF 43	2.70/ea.	Mercury. Front Hood Bumper. Made with twin retainers.
'41-'58	HF 8	2.35/ea.	Mercury, many models. Hood Bumper.
'49	HF 44	2.70/ea.	Mercury. Hood Bumper.
'50-'51	HF 20	2.50/ea.	Mercury. Hood Bumper. A perfect reproduction.
'53	HF 44-A	2.25/ea.	Mercury. Hood Side Bumpers. Four used per car.
'57-'59	HF 1	2.20/ea.	Mercury, many models. Hood Bumper.
'59-'64	HF 40	3.25/ea.	Mercury. Possibly fits additional years. Hood Bumper. Made with steel core. Held by screw.
'60-'66	SB 89-A	2.60/ea.	Mercury, many models. Bumper used for various hood, door and tailgate applications. Snap-in wire base like original.
'70-'79	HF 2	2.20/ea.	Mercury, some models, including Comets. Hood Bumper.

Engine compartments are full of small items most enthusiasts rarely think about. One such group is hood bumpers. During an in-depth engine compartment detail or restoration project, check on the condition of the hood bumpers. If your older car or truck is in need of them, Metro Moulded Parts carries a large inventory. This is just a small sample of what they offer. Courtesy Metro Moulded Parts, Inc.

of cooperation can be found with other car clubs, too. In fact, a number of clubs, especially those dedicated to vintage automobiles, circulate monthly or quarterly newsletters throughout the country. This allows members to advertise parts for sale and also ones they need.

Quality Replica Parts

Reproduction parts businesses are flourishing. Many companies are manufacturing body panels, ornaments, engine accessories, and all kinds of parts that old car restorers hunger for. But what kind of quality can you expect? Are all replicas produced exactly the same as originals?

There was a story in a well-known automotive magazine about a rather well-to-do auto enthusiast who convinced a tire maker to pull out some thirty-year-old tire molds and make new tires for his car that were exactly like the factory-originals. Expense was no object. One wonders, though, how many enthusiasts have the clout and means to accomplish such a feat?

There comes a time when you have to settle for what is available and affordable. Jerry McKee was sweating bullets one day at a concours event because his XKE had Goodyear Wingfoots instead of stock-original tires. Back in the 1970s when he bought the Wingfoots, he couldn't have cared less about stock originality. He wanted tires that looked good and would last a long time. Amazingly, the tires have lasted through to the 1990s.

Now interested in concours competitions, he worries that judging points will be deducted because his foreign car has American tires. McKee has enjoyed the beauty and self-satisfaction this car has given him for a lot of years and it wasn't until Wentworth convinced him to enter a concours event that the thought of using Wingfoots ever entered his mind. He has fun with the car and enjoys driving it. The Jaguar has never been a museum piece, rather, a means of self-expression. McKee relishes the fun times he has with it on road trips and other excursions. Fortunately, Jaguar Clubs of North America (JCNA) makes allowances for parts no longer available.

Every vintage automobile owner must reach some sort of compromise when equating the classic beauty of a special vehicle to the pleasure it can bring. We can try to preserve what we have, but the battle with nature is not always fair. Do you store a museum piece in a special garage and periodically go out and look at it, or do you drive it once in a while for the sheer thrill it brings? Do the best you can to acquire original parts to make your automobile a concours winner. But, if the ones you need are simply not in existence, don't dismay. Sooner or later, a company will devise a method to reproduce identical replicas, and concours judges will accept them. Until then, don't waste time fretting. Get on with enjoying your automobile as much as you did the first day you brought it home, even if an unoriginal replica part must be used.

Decals, Stickers, and Emblems

Over the years, some innovative car people realized that paper and plastic stickers, decals, and emblems would be in demand someday. They set out to find ways to reproduce multitudes of such items. One such company is Jim Osborn Reproductions, Incorporated, of Lawrenceville, Georgia. Their assortment of decals, stickers, and other memorabilia fills 122 pages of a catalog. (See Appendices for address.)

Decals that were originally applied to or otherwise pertained to air cleaners, valve covers, air conditioning units, emissions equipment, filters, power steering units, windshield washer parts, and others are now available in factory-original condition. This goes for Ford, Chevrolet, Chrysler, and several other auto makers, as well.

The air cleaner decal on this Corvette is plenty weathered. An identical, brand new decal would do a lot to enhance the overall appearance of this engine compartment. Decals, stickers, and emblems are available through auto dealerships, some specialty auto parts houses, and through companies that specialize in reproduction auto parts.

To find stick-on decals and other printed materials needed for your vehicle, check first with local auto dealership parts departments. Many stock those items that have been made available to them through their manufacturer. If unsuccessful, ask local car club members, look around at auto swap meets, read display and classified ads in auto magazines, or check with professional auto restorers.

Overview

Locating the right engine compartment parts for special cars and trucks is not always an easy task. Sure, an innovative enthusiast can find a part that will work, but what about the true concours buff who insists upon being the Best of Show at his or her concours event? Finding perfectly matched parts is a dilemma faced by many concours participants.

Enthusiasts are generally eager to talk cars and trucks with other folks who share their passion. Why else would car shows and events be so popular? Finding parts is as important to some as is their ability to obtain a driver's license. Therefore, take advantage of word-of-mouth communication, along with reading magazine ads and searching wrecking yards. Talking with car people is an excellent way to learn of parts that have been hidden away for decades in old barns, garages, sheds, outbuildings, and other remote areas.

For example, a serious Camaro enthusiast responded to a newspaper ad about a Camaro for sale. He dialed the telephone number and an older woman answered. He asked about the car, and the woman said it had belonged to her son who years ago was drafted into the Army and sent to Vietnam, where he had been killed. She and her husband kept the car all these years in memory of him, but were over the loss now and decided to finally sell it.

Knowing the year and particular description of the vehicle, the enthusiast asked if the car's headlights turned in when they were shut off. The woman wasn't certain, and went out to the garage to check. She returned, and said that the car didn't have any headlights. The headlights, of course, were there but were hidden behind their stock R/S shields.

This story illustrates how someone may own a special car but doesn't quite realize how rare or special it is. Or, the vehicle may have once meant a great deal to someone but no longer holds sentimental value. It could be a single automobile, or a dilapidated shed full of vintage parts and assemblies. Very often, the only way to learn of these situations is through word of mouth with friends, relatives, and their acquaintances. So, if you are in need of some unique or special auto parts, start communicating!

The old and weathered Jaguar emblem on McKee's V-12 certainly looks different than the new ones purchased at a Jaguar dealership. Once installed, new emblems make a *pleasing difference. Take your time putting emblems and decals on engine parts so that they go on straight and even the first time.*

Chapter 9

Concours d'Elegance

Concours d'Elegance events started in France decades ago. They were originally organized as a means for auto makers and clothing designers to publicly display their latest creations. Concours events were introduced to America in the late forties. Today, concours is a familiar term among avid automobile enthusiasts, not for the display of new cars and trendy clothing fashions, but for the opportunity to view some of the most uniquely designed and meticulously restored automobiles around.

A great deal of time, labor, and money has gone into the restoration of these vintage and classic automobiles. To say that many of these machines are in better condition now than they were when new is an understatement. Concours d'Elegance events are held in such places as Pebble Beach, California, Oakland University's Meadow Brook Hall in Rochester, Michigan, and hundreds of other sites across the country and around the world. Some are limited to certain automobile makes, like Jaguar, Porsche, or select American classics, while others are open to a variety of vintage and classic models from any year.

Where earlier events focused on creative new styling concepts, concours judging today is based on originality, cleanliness, and total perfection. Competition is so keen in some circles, that enthusiasts have been known to lose interest after competing against wealthy foes who hire experienced crews to com-

Concours d'Elegance events are generally held outdoors in parks, on university campuses, or other pleasantly landscaped locations. Automobiles are positioned according to their class, and visitors are allowed to look at and photograph entries. Participants frequently bring along picnic baskets full of food, and families have a great time.

119

pletely restore a vehicle during pre-event detailing. Some of these automobiles are virtual museum pieces, as they contain no fuel or fluids of any kind, thus eliminating all possibility of leaks or drip blemishes.

Fortunately, most events require winning entries to be driven up to a reviewing stand in order to accept awards. For most competitors, this is much more suitable and appropriate. After all, cars were made to be driven and it is only fitting that beautifully restored winners be operational along with being cosmetically perfect.

Maintaining Originality

One may think that maintaining originality for a newer car, say, less than twenty years old, would be easy. Not so. Some original parts have been improved over the years and it may not be easy to find replacements with identical cases, markings, or other distinguishing features found on original pieces. Such items might include batteries, distributor caps, alternators, and so on. This is where detailers must be aware of what they buy.

Dan Case has been a Jaguar concours judge since 1982 and an avid enthusiast since way before that. He notes that according to the *Judges' Instruction Manual* put out by the Jaguar Clubs of North America, Incorporated, a number of point deductions are made for such items as wrong hose clamps, improperly fitting wires, wrong battery configuration or size, exhaust manifolds painted incorrectly or displaying an inappropriate finish, throttle linkages with nonoriginal finishes, and so on. There are only a few allowances for new parts. For example, the manual states that there will be "no penalty for aftermarket electronic ignition system showing only 'black box' mounted unobtrusively." Parts can be replated but only with the same material as was applied at the factory when the car was built; for example, chrome cannot replace cadmium.

Similar rules and regulations apply to other makes and models as determined by nationally

Maintaining originality is a top requirement for most concours competitions. A boot such as this one can certainly be replaced, but the new part must be identical. Rules of concours vary a little from organization to organization. Where one may allow judging from standing positions only, another may require every square inch of every entry be intricately inspected, including the undercarriage. Some rules may also allow certain aftermarket engine part installations if newer parts are better or safer than original ones.

Jaguar concours judge Dan Case looks over the engine compartment of McKee's 1974 E-Type. He is looking for accumulations of dirt or grease, chipped paint, improper part installations, broken assemblies, and anything else that was not part of the factory-original equipment package. Notations will be made on the score sheet affixed to the clipboard he is holding and when judging is completed, point deductions will be calculated.

recognized authorities. Most of the time, national clubs like Antique Automobile Club of America, Classic Car Club of America, Porsche Club of America, and BMW Car Club of America set rules for events involving their own makes. Classes are even offered to members, designed to teach them how to become concours judges. Affiliation with national car clubs frequently brings with it monthly newsletters and tips on how to prepare for and enter concours competitions. A number of auto-related magazines such as *AutoWeek* and *Automobile* carry classified ad sections which frequently list names and addresses of national car clubs.

Strict adherence to originality is not always the theme behind every concours event, however. Many locally sponsored concours competitions are much more lenient than nationals. Some are designed just for the enjoyment of enthusiasts and offer awards for the longest distance traveled by a competitor, ladies'

choice, and so on. It gives participants a good reason to drive their special cars and an opportunity to be involved in a family-oriented event. As a judge for local events, Case takes this into consideration and realizes that in some cases, dirt or grime may account for more point deductions than lack of complete originality; such would be the situation involving the Wingfoot tires on McKee's XKE.

John Hall's 1959 Jaguar XK 150 S is in perfect condition. He has owned the car for more than twenty years. It underwent a complete restoration just a few years ago and still looks outstanding. In order to consistently win concours events, both locally and nationally sponsored, Hall has kept the car original. For him, it has been easy. This is because he bought the car with all of its factory-original equipment in place. All he has had to do is maintain it. Had the vehicle been a basket case, he may have had to do some extensive research to find and acquire original parts and accessories. In addition, his research would have had to include information about the rejuvenation of parts according to factory-original standards.

McKee's Jaguar was entered into a concours class designated for vehicles driven on a regular basis. Because of this, more emphasis may be placed on finding accumulations of dirt or grease than on locating parts that are not original. This theory is based on the assumption that vehicle roadworthiness may be best served with newer parts like a better battery or set of spark plug wires, than those offered fresh from the factory. Cleanliness, on the other hand, can be achieved with simple elbow grease, time, and patience.

The engine and engine compartment in John Hall's 1959 Jaguar XK 150 S is stunning. It was, of course, entered into a different class than McKee's. As a national champion, this car has to be perfect. Each engine compartment part is original-factory equipment. Note the cleanliness, uniform gloss, and sparkle this engine compartment exhibits.

Original equipment in a concours sense also includes such small items as decals on windshield washer fluid container lids, featured here as a circular decal on the cap. (A small amount of dust has settled on this cap after the car was driven in a slalom.)

Concours-Quality Parts Considerations

Because members share a common interest in particular makes and models, national car clubs are an excellent source of information about locating parts and materials necessary to recondition them back to original standards. It is also good to know that many reproduction parts are available through companies that specialize in such items.

Antique Automotive of San Diego, California, specializes in early Ford Parts. Their Model A catalog lists hundreds of parts for 1928–31 Fords ranging from exact battery reproductions and hood lacing to flywheel ring gears and engine compartment splash pans. They even have some NOS parts available. They rate parts according to perfection, with show reproduction quality representing as close to original as possible.

Along with offering automotive parts, companies like Antique Automotive and Jim Osborn Reproductions offer shop manuals and owners manuals dating way back. Their catalogs list which manuals are available. In addition, many specialty auto parts business representatives travel to various car shows, swap meets, and other auto events to help publicize their companies. They generally bring along plenty of catalogs and samples of their merchandise.

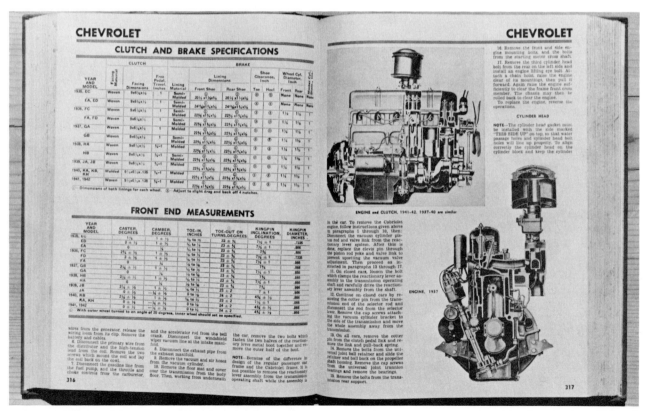

This service manual features specifications for the clutch, brake, and front end of 1937-40 Chevrolets, and is valuable to people who own cars of this era. It is through books like this that concours car owners are able to determine how specific engine parts are supposed to be installed, adjusted, and aligned. They also help owners to maintain cars in original condition.

In an effort to reproduce exact replicas of various engine parts, you will need to learn which paint colors are supposed to be applied on which parts and which ones are to remain uncoated. This is where your research into shop manuals and other literature is important. For instance, Antique Automotive's catalog lists Model A Green Engine Paint as "The exact Dark Green used on most of the cast iron parts of the engine and transmission (except exhaust manifold). Accepted by the 'Model A Judging Standards.'"

Joe Smith Automotive, Incorporated, of Atlanta, Georgia, also specializes in antique Ford parts, as well as Ford street rod parts. Ford Green engine paint is offered, along with a blue engine paint designed for 1941-48 engines. The competition among automotive specialty parts companies has greatly benefitted restoration enthusiasts. In many cases, manufacturer representatives have done their homework and discovered exactly what is needed to build parts to exact factory standards of the era, and offer paint and supplies needed for do-it-yourselfers to rejuvenate parts to concours standards.

For Jaguar detailers, XKs Unlimited of San Luis Obispo, California, publishes an in-depth catalog of available parts and also advertises their engine rebuilding service. Each engine, they claim, is finished to concours standards, acknowledging that it requires more labor to properly finish an XK engine than it does an XJ. This brings up an important question in case you decide to have someone rebuild the engine in your special automobile: Will it be finished to concours quality? If an engine rebuilder is

Porsche concours require judges to inspect vehicle undercarriages. They look for dirt, grease, grime, improper parts installations, and the like. This is part of Ridderbusch's 1979 Porsche 928 undercarriage. It is exceptionally clean, all paint work is excellent, and factory markings have been maintained.

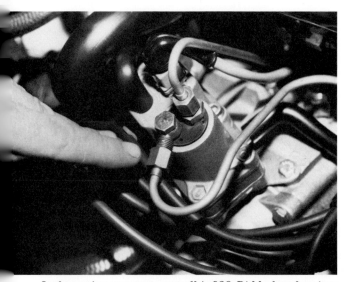

In the engine compartment of his 928, Ridderbusch points to the threads on a tube connection. Concours clean means no dirt or grease of any kind anywhere, including threads on hose connectors. Note the uniform gloss on nearby hoses and ignition wires. This is an excellent example of a national concours-winning engine compartment.

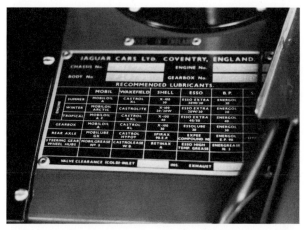

When restoring a concours contender, a great deal of care must be taken to preserve items like identification plates. Hall's is perfect, as seen here. Very light detailing efforts must be used when rejuvenating such special engine compartment parts. No harsh abrasives or potent chemicals should come in contact with them. Instead, use mild cleaners and repeated efforts with a mild paint glaze polishing product to bring out a shine and prevent damage to printing or paint.

123

unable to accomplish this for you, maybe the services of a professional restorer would be more appropriate. (See Appendices for source addresses.)

Assembling a Concours
Engine Compartment

As mentioned earlier, George Ridderbusch is a champion Concours d'Elegance veteran. He has been involved with Porsche automobiles exclusively. The attention to detail he affords his 928 is perhaps even extreme. When this car was only a few years old, he pulled the engine and completely repainted the engine compartment to make it look perfect, even to the point of attempting to match factory orange peel.

Through his years in the Porsche concours arena, he has seen competition become exceptionally keen. He points out that serious enthusiasts must read everything they can get their hands on relating to their car's year, make, and model. Some events have involved such perfect-looking cars that judges have been forced to look for imperfections as insignificant as which way cotter pins are bent. In one such event, Ridderbusch lost first-place honors and was forced into second because dust was found on the inside section of a windshield wiper blade spring.

Putting together a concours engine compartment requires a great deal of attention to detail. In addition to information already found in shop manuals and other literature, seriously consider attending concours events featuring automobiles like yours. Take lots of photos inside the engine compartments of winning entries. These will help you identify correct wiring routes, positions of hoses, and other assorted details.

Engine paint work is best accomplished while a motor is out of its host vehicle. Therefore, when it comes time to put it back in, you have to be extra careful to avoid chipping or scratching paint. One way to help avoid these problems is to mask part edges with strips of quality automotive masking tape. If need be, apply strips of thicker duct tape over masking tape for added protection. Automotive masking tape will not leave behind a glue residue, nor should it pull off paint when it is removed.

Tape placed around valve cover edges, corners of oil pans, and other externally pronounced edges on block, head, and intake manifold lends protection to those assemblies, as well as engine compartment panels. Clean, soft towels and thick cloths also work well to cushion accidental bumps while jockeying the engine into place. Apply the same type of protection to accessory parts as they are installed.

Notice the number of very small parts located throughout the engine compartment. From plastic knobs on top of the carburetors to hose clamps, there are a multitude of parts that must be cleaned, polished, or dressed in preparation for concours. The best way to approach this type of meticulous engine compartment detail is by dividing the compartment into a series of small sections. Once one section is completed, move on to another.

Unless the paint runs located next to the louvers on this hood were factory defects, they should be sanded and repaired. This is one reason why owners of concours cars should maintain small amounts of matching paint. One never knows when an accidental paint chip or scratch may occur, and the ability to quickly remedy such blemishes could mean fewer points deductions.

Tools used in the installation process also create scratch hazards. A section of garden hose placed over a ratchet handle works well to cushion accidental bumps. Large wrenches should be wrapped inside a shop towel, with only the usable end protruding. Think ahead before tackling such tasks. Would an old blanket work well to protect the firewall while working close to it? How about inner fender panels?

As much as pro street racers may like to hide wiring and other items to make it look as if their engine runs on air, concours buffs must leave everything visible that is supposed to be, including wiring. Since old wiring can cause myriad problems, especially for automobiles that have been neglected for long periods of time, restorers must consider installing brand-new wiring.

Harnesses Unlimited of Wayne, Pennsylvania, offers complete wiring harnesses for a wide variety of automobiles, some dating back as far as 1912. Although their catalog lists wiring harnesses for more than 850 different makes and models, custom harnesses can be manufactured off of schematics from shop manuals or wiring diagrams you provide.

Once an engine compartment has been outfitted with all of its parts and assemblies, work must be done to bring its overall appearance back to factory originality. This may include, but is certainly not limited to factory markings, decals, emblems,

Attention to detail goes without saying to proven concours veterans like Ridderbusch. This set of hoses and clamps demonstrate how their uniform adjustment can make this engine compartment look orderly and neat. Notice the balanced gloss presented by hoses, wires, and plastic pieces.

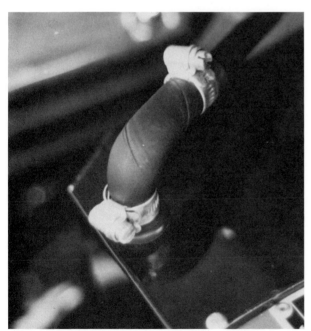

Clamps are lined up evenly and the hoses exhibit a uniform finish. This is a minor detail, of course, but it could be worth a tenth of a point deduction if it didn't look this good originally. The accumulation of several minor point deductions will eventually lead to a losing score. Concours requirements are stringent if you want to be a winner in classes like Hall's and Ridderbusch's.

Through his participation in Jaguar concours events, Hall learned this tip from a veteran. The heavy cloth cover over this radiator hose tends to fade into a light brown color after a while. A small amount of black shoe polish applied and rubbed in with a soft toothbrush brings the material back to a pleasant and uniform black finish.

Concours judges are most often enthusiastic auto buffs who enjoy the sport of concours. Generally, they have a special interest in the make of automobile they judge; in this case, Jaguar. After judging has concluded, participants frequently ask judges what measures they might take to improve their cars for the next event. A great deal can be learned from these encounters, and novice concours enthusiasts are encouraged to make such inquiries.

stickers, hood bumpers, molding, clips, connectors, and so on. Refer to shop manuals and photos of other concours winners to be sure everything is as it should be. Ridderbusch notes that some judges have even checked ignition wires to be sure they curve correctly as they come out of spark control boxes. They have also looked for scratches on the backsides of distributor cap clips, and marked down for lack of uniformity in inner fender paint gloss.

Touchup paint in the same color as parts and engine compartment body panels should be available once all engine parts have been installed. As hard as you try, accidents are bound to happen and you will surely need to touch up a few paint chips or nicks. Use an artist's fine paintbrush to slowly build up paint inside chips. Wait at least an hour before applying second or succeeding coats to give paint time to dry. Should chips go down to bare metal, primer will have to be applied first.

After about a week to ten days, tightly mask off the small chip area and use fine sandpaper (1200 grit) to gently smooth the paint spot until it is even with its surrounding surface. Remove masking tape and polish that area with a mild glaze. Do not apply wax for a couple of weeks so paint will have plenty of time to dry. (This allows paint solvents to completely evaporate instead of being sealed in by a wax product.)

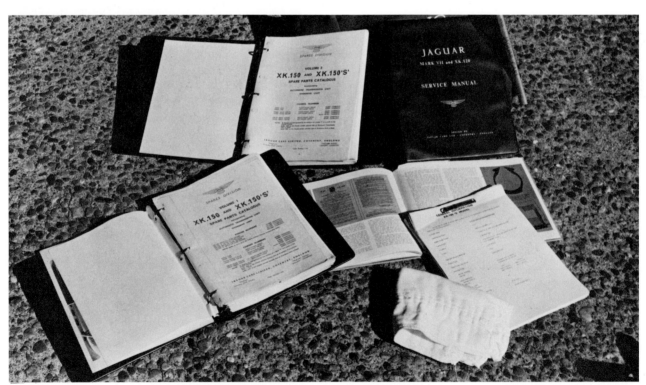

An assortment of Jaguar literature that Hall brings along to concours events. If a judge has a question pertaining to the way a specific item is addressed in his car, he can refer to official Jaguar information for proof that he has conformed to Jaguar standards. This is typical of most serious concours contenders. They have spent a great deal of time researching the various aspects of their concours car so that repairs or restoration efforts are done according to factory specifications.

Perfect Gloss and Balanced Uniformity

A new concours engine compartment is not complete nor ready for judging until all of the little details are attended to. Things like hose clamps, wire routes, paint gloss, rubber parts gloss, and minor parts adjustments may be considered trivial but are certain point deductions if noticed by a serious judge.

Ridderbusch's Porsche 928 features a set of rubber hoses on top of the engine. Each hose is equipped with a hose clamp. In order to balance the appearance of these clamps, he adjusted each one so its screw mechanism was in the same relative position as the others. A minor item, surely, but one that could save tenths of a point in concours. And yes, the difference between first and second places has been decided by only a tenth of a point in more than one instance.

Wiring must be routed according to original factory designs. This includes curvature, boot positions, connection angles, and the like. In addition, the type of tip attached to wire ends is of equal importance. Wires should not be blemished with any paint overspray and their gloss should be uniform and subtle—super clean without hints of shiny dressing overapplication.

High gloss on anything is not necessarily good for concours. This is where rookie concours participants frequently make mistakes. They expect concours to be similar to regular car shows where entries like hot rods, street rods, and customs are highly polished with lots of shiny metal and brilliant, glossy finishes. Not so in concours. Gloss must be uniform, clean, and free from any unoriginal mirror-like sheens. In other words, they must be clean and polished without the glitz; sanitary; rich but not necessarily eye-catching; bright but not brilliant. Subtle is the key for concours, with parts strikingly perfect in that everything has identical degrees of gloss.

The same goes for paint, wiring, and rubber parts. Ridderbusch, McKee, and Case have had best concours paint polishing results with Meguiar's #7. Apply this glaze in straight back-and-forth movements to avoid unnecessary swirls. Parts or body sections may be lightly polished for hours until just the right degree of gloss is reached.

Hoses and wires receive very light buffing with dressing. Ridderbusch and Hall like Meguiar's #40,

All entries in this concours event went through a mechanical check before actual judging began. Here, a participant goes down a check sheet that lists each mechanical item the car's owner must operate. It is a pass-fail measure encompassing such things as lights, horn, windshield wipers, and the like.

The detail kit that Hall brings along to concours events. It contains important parts of his Jaguar library, a plastic bag full of clean, soft cloth baby diapers, an assortment of mild polish and glaze products, wide masking tape used to pull lint off of the roof and carpet, some soft toothbrushes, an assortment of artist's paintbrushes, and vials of touchup paint. This is a typical example of what most concours competitors bring with them to events.

McKee likes No Touch, and Wentworth prefers Armor All. The brand of dressing you use is not nearly as important as its application. All excess dressing must be buffed off. In fact, using too little is better than using too much. Apply just a dab to a clean cloth and work dressing into hoses and wires. When a uniform gloss has been achieved, continue to buff until that sheen is dulled to almost no shine whatsoever.

The best way to understand subtle, balanced, and uniform gloss is to go to a concours event and see for yourself. At first, you may be surprised at the lack of gloss. You may even be disappointed when you initially see a winning engine compartment. It will not jump out at you with lots of glitz and glamor. After a few minutes of examining the detail work, however, you will grow to respect such perfection. Everything will be in its place without a speck of dust or dirt on it.

Balanced uniformity essentially means that nothing in the engine compartment stands out more than anything else. All parts look great, super clean, and perfectly matched, as if they are all related. Parts in a winning concours engine compartment should never look out of place, newer, or older than the others. In addition, everything must work, even the horn.

Concours Tips

To prepare a vehicle for concours competition, Ridderbusch uses Gunk engine degreaser sparingly to remove any unusually heavy concentration of grease or grime. Dab some on a cloth to loosen up and wipe away smudges. He suggests treating paint surfaces as you would your skin, with gentleness and consideration for the long run. A potent chemical may do a marvelous job of cleaning quickly, but what has it done to the surface paint or polished finish it was used on? Case feels the same way and believes too much degreaser may tend to soften hoses. Instead, he relies on Simple Green to clean extra-dirty surfaces, even if it takes repeated applications.

From a judge's standpoint, Case believes properly prepared entries are easiest to judge. Cars with a number of problems create confusion when it comes time to figure out deduction points and calculate scores. A perfect car is easier to score because there will only be a few things to look for.

Initially, Case will walk around an engine compartment to get a general picture of the vehicle and how well it has been prepared. Then, he will start at one corner and work his way around, inspecting every square inch along the way. He will look for signs of rust, dirt, debris, and flaking paint, and deduct points as needed. Points will also be deducted for items that are overdetailed.

Overdetailing refers to things like paint that's too shiny, aluminum polished to look like chrome, hoses too glossy, and so on. If a judge sees a part he or she believes is overdetailed, the contestant will be asked to show in which official Jaguar publication overdetailing is original for the part in question.

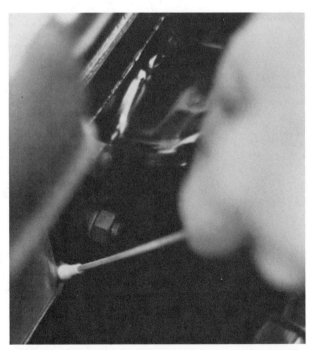

In bright sunlight, small paint chips are quite noticeable, as are other minor flaws. Here, a concours competitor uses an artist's paintbrush to touch up tiny paint nicks along the firewall of his car. Be sure to bring along a small bottle of thinner for cleaning brushes and any paint that may have inadvertently gotten on your fingers.

Hall uses a cotton swab attached to a long, wooden handle, like those seen in doctor's offices, to remove hints of oil from along an engine gasket edge. This is truly attention to detail and something he takes in concours stride. These types of cotton swabs are great for reaching into extra-tight spaces to remove traces of dirt or oil just before judging time. They can be purchased at pharmacy counters and medical supply stores.

JAGUAR CLUBS OF NORTH AMERICA, INC.
555 MacArthur Blvd., Mahwah, NJ 07430-2327
Telephone: (201) 818-8500

REV. 1991

DATE _____

HOST CLUB & LOCATION: _____ CAR ENTRY NO.: _____

YEAR, MODEL & STYLE OF ENTRY: _____ CLASS: _____

ENTRANT'S NAME: _____ JCNA NO. _____ CLUB MEMBER OF: _____

ADDRESS: _____ JUDGE'S NAME & NO. _____

CITY _____ STATE _____ ZIP _____ OWNER OR FAMILY MEMBER PRESENT ☐

Deductions for non-authenticity will be considered invalid unless the owner or family member is informed that the deduction is being taken, and is given the opportunity to prove authenticity of the item.

IN ORDER TO ENSURE CORRECT RECORDING OF SCORES, PLEASE FURNISH ALL INFORMATION

Judge's Note: When judging class 15: Competition Cars, see section D, item 1 on p.11 of the rule book.

EXTERIOR

BODY, DOORS, BONNET, etc., (includes, door jambs, sills, shutface, hingeface and rubber seals)

	MIN. POINTS PER DEFECT	MAXIMUM TOTAL DEDUCTION DHC / OTS	MAXIMUM TOTAL DEDUCTION SEDAN / FHC	TOTAL
Dent/ripple	.1	6	8	.
Poor repair	.1	6	8	.
Poor fit	.5	6	8	.
Crack	.5	6	7	.
Rust	.1	6	7	.
Poor rubber/deteriorated	.5	10	12	.
Non-authentic see guide		24	30	.

PAINT FINISH

	MIN. POINTS PER DEFECT	DHC/OTS	SEDAN/FHC	TOTAL
Scratch	.1	6	7	.
Chip	.1	6	7	.
Peel	.1	6	7	.
Fading (Obvious)	2	5	7	.
Worn paint	1	6	7	.
Orange peel, fisheye/etc.	.5	5	7	.
Paint overspray	.1	5	7	.
Non-authentic see guide		23	29	.
Cleanliness		21	26	.

TIRES

	MIN. POINTS PER DEFECT	DHC/OTS	SEDAN/FHC	TOTAL
Cracked/crazed	1	4	4	.
Less than legal tread	1	4	4	.
Non-authentic		5	5	.
Cleanliness		8	8	.

GLASS

	MIN. POINTS PER DEFECT	DHC/OTS	SEDAN/FHC	TOTAL
Discolored/cloudy	.5	10	10	.
Scratches/chips	.1	8	8	.
Cracked/split	.5	8	8	.
Non-authentic see guide		16	16	.
Cleanliness		15	15	.

TOP, SIDE CURTAINS TONNEAU & BOOT COVER (excluding glass)

	MIN. POINTS PER DEFECT	MAXIMUM TOTAL DEDUCTION DHC / OTS	MAXIMUM TOTAL DEDUCTION SEDAN / FHC	TOTAL
Scratch, tear or hole	.1	4		.
Poor fit	.5	4		.
Frayed/loose bindings	.1	4		.
Faded	.5	4		.
Creases/wrinkles	.5	4		.
Non-authentic see guide		12		.
Cleanliness		5		.

CHROMEWORK (including accessories, tailpipes and resonators)

	MIN. POINTS PER DEFECT	DHC/OTS	SEDAN/FHC	TOTAL
Dents/ripples	.1	7	7	.
Pits/rust	.1	7	7	.
Lifting/peeling	.5	7	7	.
Scratches, worn, faded	.2	7	7	.
Paint overspray	.1	7	7	.
Poor rubber component	.5	7	7	.
Non-authentic see guide		25	25	.
Cleanliness		20	20	.

WHEELS

	MIN. POINTS PER DEFECT	DHC/OTS	SEDAN/FHC	TOTAL
Damaged wheel/spoke	.5	10	10	.
Pitted chrome/paint chip	.1	10	10	.
Rust on wheel/spoke	.1	10	10	.
Non-authentic see guide		18	18	.
Cleanliness		20	20	.

EXTERIOR: MINUS DEDUCTIONS OF [____]

BOOT

PAINT, SIDE PANELS, MAT or CARPET

	MIN. POINTS PER DEFECT	MAX. TOTAL DEDUCT	TOTAL
Chips, scratches/etc.	.1	7	.
Poor paint, repair/dents	.1	7	.
Scratches/tear in mat	.1	7	.
Pits, rust/corrosion	.1	7	.
Non-authentic see guide		17	.
Cleanliness		26	.

TOOL KIT

	MIN. POINTS PER DEFECT	MAX. TOTAL DEDUCT	TOTAL
Scratches, dents/rust	.1	7	.
Stained, torn/faded manual	.5	5	.
Torn/faded tool pouch/liner	.5	5	.
Non-authentic see guide		10	.
Cleanliness		5	.

SPARE TIRE, WHEEL & COVER

	MIN. POINTS PER DEFECT	MAX. TOTAL DEDUCT	TOTAL
Damaged wheel/spoke	.5	2	.
Pitted chrome paint chip	.1	3	.
Rust on wheel/spoke	.1	2	.
Cracked/crazed sidewall	.5	2	.
Less than legal tread	1	2	.
Non-authentic see guide		6	.
Cleanliness		7	.

EXTERIOR: MINUS DEDUCTIONS OF [____]

SIMAX, NY

Here is a sample concours scoring sheet from the Jaguar Clubs of North America (JCNA). Judges inspecting a vehicle will follow such a scoring sheet's guidelines about minimum and maximum deductions for defects. To learn how your specific marque's club or association judges its concours events, request a sample scoring sheet along with complete rules. Courtesy JCNA

Case says that participating in the local chapter of the Jaguar Clubs of North America was the best way for him to learn about Jaguars. Not only was he able to learn from other members, but a lot of information was received from the Jaguar factory, as it sponsors JCNA.

When judging had been completed at events, Case would talk with judges to learn why points were deducted for particular items and what he needed to do to correct them. He notes that members of JCNA are urged to participate in concours events. Like most other national car clubs, members are trained to become judges. They learn judging skills through classroom sessions and live training with cars onsite. Members must pass tests before they are qualified to judge competitions.

Case recommends that serious auto enthusiasts attempt to build libraries of books that pertain to their automobiles of interest. Books with photos of original makes and models are most valuable, as they show how items are supposed to look and be positioned according to the factory. For Jaguar owners, a copy of the JCNA *Judges' Instruction Manual* would be handy.

Before judging commences on the appearance and cosmetic aspects of a vehicle, its mechanical operations are tested. All items must work, such as windshield wipers, horn(s), headlights, taillights, license plate lights, turn signals, and so on. Operation or failure of such is noted on a check sheet. Throughout the *Judges' Instruction Manual*, reference is made to authenticity in color, plating, and original-equipment parts. Unless original parts are no longer available, points are deducted for nonfactory replacements.

As far as cleanliness is concerned, the following passage from JCNA's *Judges' Instruction Manual* says it all:
"D. Engine Compartment
Despite the difficulty of maintaining an engine compartment, cleanliness and condition of painted and other surfaces of the engine compartment of a car prepared for concours should be excellent. NO EXCUSES."

Concours d'Elegance contestants are given an allotted time before judging to accomplish last-minute cleaning details. Time limits vary from event to event. When "Rags Down" is sounded, all detailing work must stop and judging begins. To stay on top of their cars' engine compartment condition, avid concours enthusiasts clean them after every drive. Once parked in position at an event, soft clean cloths, polish, and touchup paint come out and are not put down until a "Rags Down" warning is heard about five minutes before the official "Rags Down."

Serious competitors usually bring along special concours detailing kits put together at home (along with picnic baskets and refreshments). These include items like baby food jars filled with different touchup paint colors and thinner, artist's paintbrushes, towels and cloths, paint polish, chrome polish, soft toothbrushes, soft paintbrushes, and so on. They will then watch judges carefully and make mental notes of areas or parts that receive most of their attention. When judging has been completed, contestants frequently talk with judges to gain insight on how their cars can be improved. Written notes are normally made on judging sheets so contestants are able to determine where points were deducted.

Local concours events are generally much more relaxed than national ones, although competition is keen among regional neighbors. Ridderbusch suggests anyone with an interest in concours simply enter one. The worst one can do is come in last place. "But," he says, "look at all the fun you'll have and interesting cars to look at and knowledgeable people you'll meet." Case agrees and encourages enthusiasts to get involved with car clubs of the same make or model car. "You will meet some very nice people and learn more than you thought possible," says Case.

This advice appears to hold true for most serious auto enthusiasts. In bygone decades, car clubs may not have been construed as anything more than a bunch of kids running around in fast cars that made a lot of noise. Whether or not that was the case, car clubs today generally include members from all ages, backgrounds, and interests. Enthusiasts who have learned the hard way through novice experiences are now eager to show new members tricks of the trade and share learning experiences of all kinds.

Ridderbusch is a prime example. It took him nine years to restore his 1957 Porsche 356 A. When it was completed in 1973, he entered his first concours competition. Expecting to win, he was quite surprised at his fourteenth-place ranking out of seventeen entries in his class. With well over eighty concours awards now decorating a wall in his home, he is grateful for all he learned through friends and acquaintances he met while involved with the Porsche Club of America. Among his peers today, he is regarded as a local Porsche expert and they are elated when he conducts restoration classes.

Individuals like Ridderbusch belong to hundreds of car clubs around the country. Their interests range from antiques to hot rods and pro street machines to serious concours competition. Once you get your car to baseline, knowledge and expertise provided by other car club members who might also be car show veterans and concours winners could prove quite valuable.

Chapter 10

Complete Restorations

Absolute perfection is required for restoration of a future concours national champion. Although concours had not entered John Hall's mind until restoration of his 1959 Jaguar 150 S was completed in 1988, the car was put together in better-than-new condition simply because he wanted it that way. It was only after realizing that his car looked better than a lot of others which had already won concours that he decided to join the Jaguar Clubs of North America and give concours a try.

This may not be the case with every car owner seeking to restore the engine compartment of his or her automobile. There are varying degrees of restoration, all somewhat governed by a vehicle's expected function, its value, and the restorer's overall expertise, time, and money limits. This does not mean that an inexpensive restoration has to look amateurish, but that the strict adherence to factory-original parts and specifications may have to be loosened in lieu of aftermarket parts and accessories.

Complete engine compartment restorations will include a lot more than just engine and parts reconditioning. Bodywork and painting problems must be addressed, as well as suspension and steering concerns. Realize your own limitations before tackling steering, brake, and front suspension reconditioning. These assemblies are crucial to safe vehicle operation and just a small flaw in their installation could result in tragedy.

What to Expect

Just about any experienced auto restorer will admit that restoration projects generally take two to three times longer to complete and at least two to three times more money than initially estimated. This phenomenon is not due to one's haphazard work methods, but results from the inevitable unexpected problems that were not visible at the start of the project. Problems hidden underneath coats of paint, motor parts, and other obstacles may go unnoticed until dismantling uncovers them.

Rust is a major problem, especially in areas around front fenders on some particular makes and models. Older American cars and trucks were frequently designed with large front fenders that included inner pockets prone to accumulate dirt, road grime, and other debris. Left unattended for extended periods, these masses were able to trap and hold in moisture, giving rust and corrosion a significant start.

This pristine Jaguar engine, complete with a porcelain exhaust manifold and polished aluminum engine pieces, was restored to original factory specifications because John Hall wanted it that way. As things turned out, he later got involved with Jaguar Clubs of North America and became a concours winner.

As dismantling progresses to the removal of inner fenders and other sheet metal work in engine compartments, restorers are often faced with rust problems that had not been initially factored into restoration projects.

As you go about taking parts off of a firewall or inner fender component, you may begin to notice rust problems around mounting holes, along seams, inside sheet metal joints, and under components like the battery tray, radiator supports, and weatherstripping. The amount of rust found is significant. If it is just a touch around mounting holes, repair work should be minimal. But, if seams show corrosion along their entire length, you may have to rely on extensive metalwork for repair or the purchase of a new sheet metal panel.

Don't be surprised if your older project vehicle presents you with a lot more rust problems than initially perceived. Many experienced auto restorers have been shocked by the amount of rust that was hidden under painted surfaces, and at the large amount of metal corrosion that had actually taken place. In many cases, rust started on the backside of sheet metal parts and hardly showed through to the front because heavy exterior paint layers somehow stayed intact.

Once you start actual restoration work and begin reconditioning parts and assemblies, you will become more aware of and attentive to details. As one item nears completion, you will more than likely notice that another does not quite meet the same quality standards. It happens all the time. Before long, you might find yourself going back to some of the parts you finished early in order to make them cleaner, brighter, smoother, or more intricately detailed.

This is a common occurrence for novice do-it-yourself restorers because they learn so much along the way. Eager to get started, they might clean, sand, and paint the alternator or generator brackets first and not learn about epoxy primers or specific engine compartment paint gloss percentages until after discussing the project with an experienced car club member or friend. The more they learn, the better job they'll do and as the work progresses, they realize that the parts worked on earlier do not look nearly as good in comparison to those rejuvenated late in the game. Hence, conscientious restorers redo those items to bring them up to their new quality standards.

Expect your workplace to soon become cluttered with lots of parts. Since an engine compartment must be empty before work can begin, all of the items

Front suspension, steering, frame, and brake restorations must be flawless in order for vehicles to be safe driving machines. Unless you have experience working on these kinds of components, have a professional assemble, adjust, and align them. If you did complete restoration work to these systems yourself, at least have them checked out by a professional before taking the vehicle out on the open road.

The engine compartment was painstakingly restored along with the rest of this car. Since the vehicle was stock to begin with, restoration was simplified; all original parts were in place, thus eliminating the search for new ones. Note the crisp condition of wires and grommets along the firewall.

taken out must be stored somewhere. It's a good idea to store related parts in common containers to make reconditioning and eventual installation easier. If you don't use this type method, you will have a terrible time finding all of the little screws, washers, nuts, bolts, clips, and fasteners when you need them.

Parts storage is a major concern and should not be taken lightly. The vast majority of basket case cars, trucks, and motorcycles found at swap meets, garage sales, and in classified ads probably all started out as restorable machines but ended up in disaster because an overeager restorer failed to maintain any semblance of order. A definitive plan of action that includes short-term goals and manageable parts storage should result in a job well done and in a timely fashion.

Since engine compartment restoration entails a great many tasks, you must have access to shop manuals that pertain to the year, make, and model of your automobile. One cannot expect to install all engine compartment parts with precision when torque specifications, wiring routes, and other pertinent information is not available. You should have this material on hand before a project begins, not only for installation references but also for determining how certain parts should be dismantled.

If you have never undertaken an automotive restoration project the size of the one you are about to embark upon, heed the advice of many veterans and talk to people who have experience restoring engine compartments similar to yours. They should know where to locate specific parts and about how long it takes to get them; how some unique items may be mounted or supported; the best way to remove parts without causing unnecessary damage; which machine shops provide the best quality of work; and lots of other helpful tips. Having a good idea of what lies ahead will enable you to plan out a course of action and complete your restoration with few surprises.

Overall Options

Engine compartment restoration can take on a number of different faces: attention to originality for concours; slight modification for the ultimate installation of a bigger motor; some custom work for a street rod or hot rod look; or a complete facelift en route to a total pro street appearance. The kind of automobile being restored and the interests of the owner have a lot to do with which options are realistic.

Car body and frame ready for an engine, tranny, and steering controls. Taking a car down to this sort of frame-off restoration will require a tremendous amount of time and effort. Each piece will have to be cleaned, reconditioned, painted, and detailed. Expect the unexpected, and realize that major projects like auto restorations require a wide range of mechanical skills.

To estimate the amount of sheet metal work required to bring an engine compartment's appearance up to your expectations, consider using a Spot Rot gauge to measure how much rust is present on panels. If repainting is a concern, use a Pro Gauge to measure how much paint is already covering sheet metal. This information can be helpful when determining how much work will be needed to complete engine compartment bodywork and/or painting. Tools courtesy Pro Motorcar Products, Inc.

An engine compartment being prepared for show-car quality and the installation of a fresh V-6 engine. Sanding marks are signs of bodywork repair. As your bodywork skills improve with practice, don't be surprised if you find yourself going back to areas where your work first began and attempting to rectify mistakes. This kind of labor-intensive work will probably not be repeated on the same automobile for a very long time, so have the patience to do it correctly the first time around.

Parts tags are necessary items that should be used frequently during any restoration project. On the tag, indicate names of parts, where they came from, and include any helpful hints for reinstallation. Tags courtesy The Eastwood Company

Hall could never imagine his Jaguar XK 150 S set up to look like a hot rod, nor would Mycon like his 1948 Chevy as much if it were simply stock. Both enthusiasts have different automotive interests and their cars represent their varied preferences. So, before you start cutting sheet metal panels or tossing away engine accessories, have a definite idea of what you want your car or truck to look like when it is all put back together. Plenty of ideas are available through car shows, car club events, and auto magazines.

Mycon has gone to a lot of work and expense turning his stock 1948 Chevy into a street rod. A sheet metal panel was welded onto the firewall to flatten that area out, and a new front-end suspension and steering system were installed to accommodate a V-8 engine, power steering with a tilt wheel, and disc brakes. The car's entire nose was dismantled to allow for complete bodywork repairs and new paint. He painted the engine black and the valve covers Mycon Purple to match the car body. Plenty of billet accessories were included to finish off his restoration-modification package.

Many times, an engine compartment restoration takes on a new face with the installation of a newer and bigger motor. High-performance shops and a number of specialty auto parts outlets carry specific motor mounts and other parts to make engine conversions work smoothly. You must do your homework first, though, to be sure your new engine will easily fit into the car you are restoring, and that the necessary conversion parts are readily available.

Along with frame and suspension changes, Mycon learned that other modifications had to be made for all of the featured accessories to function properly. Such changes included a shield between heater hoses and headers, a tapped heater hose connection into the intake manifold to make way for an air conditioning compressor, mounts for miscellaneous engine compartment accessories, and custom tube bending for brake and fuel lines.

The problems Mycon faced during restoration and modification are typical of such projects. Unless a person does this type of work on a consistent basis, it is almost impossible to foresee trouble spots before you actually have to deal with them. That is why jobs like this generally take much longer to complete than originally anticipated.

Mycon is fortunate that the engine compartment in his Chevy was big enough to sport a V-8 engine. Other vehicles, like 1930s vintage Fords, do not always offer that same kind of roomy convenience. Serious car restorers and builders take such obstacles in stride and adapt by employing innovative ideas. For example, a restorer might install an electric fan to assist in cooling a V-8 engine that could use a

Putting nonstock motors into older cars will most often leave restorers scratching their heads trying to figure out how certain parts are supposed to fit into specific areas. Mycon did this while attempting to fit an air conditioning compressor on the front of this engine, while at the same time trying to route a radiator hose and heater hose to the same general area. He had to tap a line into the base of the thermostat mount for a heater hose and ended up using a swivel thermostat housing to accommodate the radiator hose.

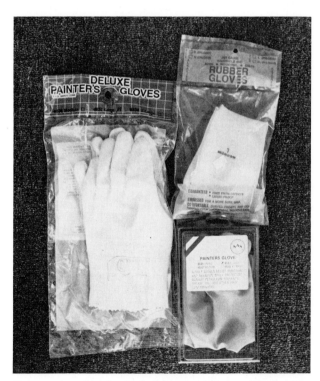

Protect your hands against potent paint and body filler materials by wearing quality painter's gloves. Lacquer thinner, paint reducer, and other similar materials quickly remove natural skin oils to leave hands dry and prone to cracking and bleeding. In addition, continued exposure to harsh chemicals can cause skin rashes and other related health problems. Painter's gloves are widely available at autobody paint and supply stores and some auto parts houses.

Any chores associated with bodywork, sanding, or painting require users to wear respiratory protection. Filter masks are fine for sanding work, but you will need the added protection of a quality respirator when using primers and other paint products. Be sure you read product labels to determine what type of protection is recommended.

radiator larger than those made to fit. A change in fan design may help sometimes, too.

Time and money could have been saved if Mycon would have simply refurbished those parts that were already on his car, such as the front-end suspension, engine, steering, and so forth. But, the result would have been something completely different than desired. The car was in neglected condition, didn't run, and needed extensive body and paint work. Whichever option he took, stock versus street rod, he knew that lots of work and financial commitment was ahead of him.

You may have to make the same kind of choice with your car. Is it going to be most valuable as a stock machine, possibly a concours contender? Or, is it something that could look extra special as a street rod, pro street machine, or wild custom creation? Should it remain stock but altered just enough to fit in a big V-8? Or, just fixed up as it is and then driven? By all means, should you decide to modify or customize the engine and engine compartment on your project vehicle, get accurate cost estimates on the parts first. Many companies offer aftermarket parts for custom engine installations and you should be able to get a fairly close estimate before work starts.

Bodywork Tools and Materials
Completely restoring an engine compartment will most likely require the firewall and inner fenders

Bodywork restoration frequently calls for welding to repair old welds, install patch panels, or fill holes left behind by the removal of old assemblies. Low-amperage welding works best on sheet metal. Too much heat will simply cause metal to disintegrate and make welding impossible. This machine works off of 110volt household current and is perfect for small welding jobs on materials up to ³/₁₆in thick. Welder, gloves, and helmet courtesy The Eastwood Company

be taken down to bare metal. This task can be accomplished with a chemical paint stripper, high-speed rotary sander, dual action (DA) sander, or sandblaster.

Chemical paint strippers do a good job and are generally easy on underlying metal. However, you must wear heavy-duty rubber gloves and eye protection. If a product label instructs users to wear respiratory protection, be sure you do that as well. To prevent unwanted scratches, use a plastic squeegee to scrape off paint rather than a putty knife or other hard object. Chemical paint stripping is messy, and a

A high-speed rotary sander with a 36 grit sandpaper disc is used to remove paint from an autobody sheet metal surface. This tool and disc combination works fast to remove paint, rust, and other contaminants from body surfaces. Users should wear a full face shield or goggles while operating tools like this. Tool courtesy The Eastwood Company

The Welder's Helper is used to back up small holes, rips, and voids while attempting to weld them closed. The paddle is made of copper and is attached to a covered 6in aluminum nonslip handle. The copper paddle is not affected by welding bonds while supporting work from the rear. Photo courtesy The Eastwood Company

means for catching gooey residue must be provided. Pushing or parking a vehicle on top of a large sheet of heavy plastic is one way. Once stripping is complete, fold up the plastic sheet and then discard it according to recommendations from your local waste facility or hazardous waste disposal company.

High-speed rotary sanders do a great job of removing paint, especially when outfitted with coarse sandpaper. This job is also messy, as it creates clouds of sanding dust. Consider wearing goggles and always wear a quality filter mask or respirator. Mycon has had good luck using a high-speed rotary sander and 36 grit sandpaper to remove accumulations of paint, body filler, and rust. This method often leaves behind heavy sanding scratches, however, and may be unnecessary for engine compartments that do not suffer rust or body filler problems.

DA sanders with 80 grit sandpaper may take longer to strip paint than a high-speed rotary model and 36 grit, but sanding scratches left behind will be much less defined. You will have to experiment on your engine compartment panels to see which works best for you.

Sandblasting works well, but poses problems with regard to dusty residue and sheet metal warping damage. Sandblasting media will find its way into suspension pieces, brake assemblies, and a host of other areas where it is not wanted. You can mask off sections to protect them from residue accumulation but still run a risk of body damage due to the sheer

This transmission has been sprayed with DP 40 Epoxy Primer in preparation for a full paint job. The same procedure is undertaken for engines that have been removed from their vehicles, after they have been prepped and masked. This is an excellent way to prevent rust and corrosion from undermining an otherwise great paint job.

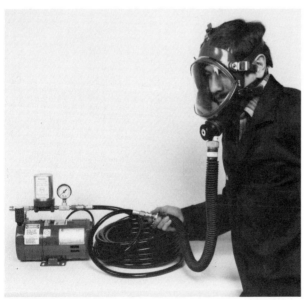

The catalyst used in urethane paint products contains isocyanate material. Users are strongly advised to wear fresh air respiratory protection whenever spraying such material. Units can be rented from rental yards or purchased through autobody paint and supply stores or mail-order tool and equipment dealers. Photo courtesy The Eastwood Company

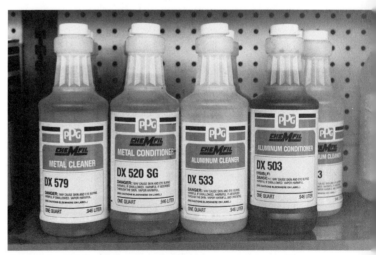

A complete PPG paint system. All products are compatible with each other. Shown here, from left, are epoxy primer, lacquer thinner, primer-surfacer, sealer, and a three-step paint product which includes paint, reducer, and catalyst. Regardless of the paint brand you use, be sure all products are part of a complete system to ensure compatibility and satisfactory results.

Corroless is a rust treatment product that changes rust accumulations into a magnetite to stabilize corrosion. In addition, it contains millions of heat-hardened, self-leafing glass flakes that form a tough, impermeable finish to seal out moisture and other corrosive elements. Once rust has been successfully treated with this product, paint can be applied over it with no problem. Photo courtesy The Eastwood Company

force behind sandblasting operations. Mycon saves sandblasting work for other assemblies that can be blasted separately, away from vehicles.

The pressure and type of media used for sandblasting body panels is very important. Media too coarse or pressures too high will quickly warp or pit sheet metal panels. Operate sandblasting nozzles at a 45deg angle to reduce warping, and constantly check the progress of your work. Fully explain your sandblasting needs to an autobody paint and supply jobber or sandblasting supply store representative before purchasing material. Most businesses that carry such supplies also display charts that list recommended media and pressure combinations for use on specific work surfaces.

Once paint is stripped, you will have to repair any body damage discovered. The Eastwood Company carries a wide selection of body repair tools and equipment such as hammers, dollies, welders, slide

Plastic body fillers are applied over sheet metal repairs and then sanded smooth to leave behind a finish suitable for painting. If your car or truck has suffered engine compartment body damage and hot metal repair work, like leading-in or panel beating, is not a reasonable alternative, you will have to use a plastic body filler.

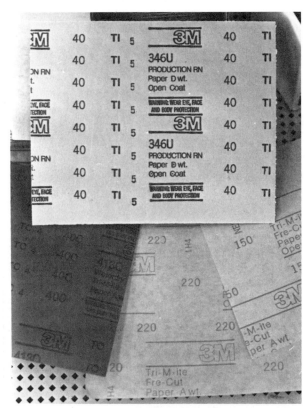

Sandpaper grit is designated by number. The lower numbers refer to the coarsest grit while higher numbers represent ultra-fine grits. For example, 36 grit is very coarse and 1200 grit is very fine. An assortment of grits will be needed to accomplish paint removal, body filler, and primer-surfacer sanding. Confer with an autobody paint and supply jobber to determine how much of each grit should be needed for the job you are contemplating. To ensure accurate sanding results, always use a sanding block instead of your hand. Sanding blocks are available in various sizes and shapes. Their flat surface allows user to maintain equal pressure with each sanding stroke.

hammers, dent pullers, and so on. Use tools like these to straighten body parts as close to perfect as possible—at least to within ¼in. Panel beating, heat shrinking, and leading-in procedures can be successfully accomplished on older vehicles that sport thicker sheet metal engine compartment panels. Newer vehicles with thinner metal may have to settle for coats of body filler that will be sanded smooth.

Rust problems must be corrected before work can continue. Use a high-speed rotary sander and coarse sandpaper to remove all traces of rust. Then, treat metal surfaces with a chemical rust remover like Oxi-Solv, Captain Lee's Spray Rust Away, or other similar products. Be sure to follow label instructions, including user safety recommendations. Seriously consider applying rustproofing materials both during and after paint work has been done.

Once bodywork and rust elimination have been completed, plan to spray engine compartment panels with an epoxy primer like PPG's DP 40. Epoxy primers are waterproof and do an outstanding job of helping to prevent future rust and corrosion. After allowing for sufficient drying time, coats of primer-surfacer should be sprayed onto panels. This high-solids primer material fills in minor low spots, like sand scratches, to enable restorers to achieve perfectly smooth panel surfaces.

The kind of paint used for your engine compartment depends on your painting experience and equipment. Acrylic lacquer is good for novices be-cause minor imperfections can be sanded smooth and then repainted. Lacquer dries fast and is very forgiving. Acrylic enamels cover with just a few coats but cannot withstand any sanding. Urethane paint products are very durable and feature excellent coverage with just a few coats and can also be sanded should nibs blemish a surface. However, urethanes require a catalyst hardener which contains isocyanate material that should not be inhaled. Product manufacturers recommend users wear a fresh air respirator whenever spraying urethanes with such hardeners.

The products used for body filler operations, primer, and painting are available at autobody paint and supply stores. You must use products that are entirely compatible with each other. The term paint system refers to all paint products used through an automotive paint process. These include primer materials, thinner/reducers, wax and grease removers, flexible additives, fisheye eliminators, primer-surfacers, sealers, adhesion promoters, and paint. Mycon recommends that you consult with an autobody paint and supply jobber before purchasing paint products. Explain the kind of job you plan to do and the results expected. Be frank and up front. These materials will

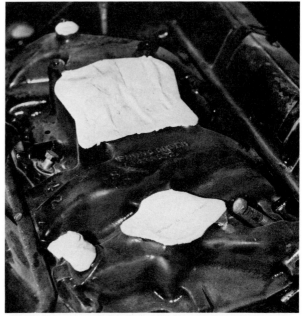

An engine being sanded and prepped for painting. Masking is an important consideration whenever paint work is involved. All engine surfaces that will be covered by a gasket should be masked to eliminate paint build-up which could cause gasket failure and result in fluid leaks. Lay wide pieces of masking tape over irregularly shaped items and then trim excess tape with a sharp razor knife or blade. Autobody paint and supply stores carry informational material on the paint products they sell. This material includes mixing instructions, drying times, application instructions, and compatible additives.

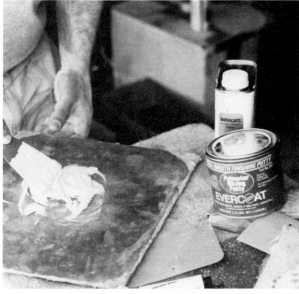

Although body fillers strengthen repair spots when applied properly, their surfaces are not always perfectly smooth. Many professional autobody technicians cover filler materials with a thin coat of glazing putty. This fine material is not strong enough to offer dent repair capabilities, but does an excellent job of filling pinholes and sand scratches in preparation for paint work.

The type of paint you spray inside the engine compartment of your special car or truck should be suitable for the type of service you expect from the vehicle. Custom candy and pearl paint jobs are not always easy or possible to touch up, whereas other kinds of paint systems easily accept chip and nick touchups. Special paint jobs should be saved for show cars that will not be subjected to road hazards and other conditions where paint chips or nicks are likely to occur.

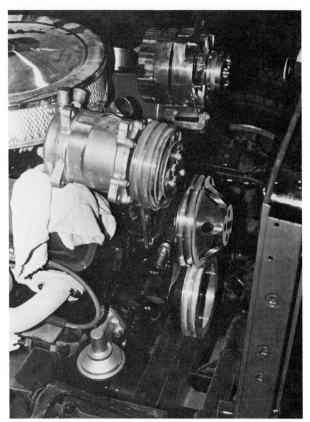

When you start putting parts back on your engine and in the engine compartment, take time to make sure they don't show any paint chips or scratches. A clean, soft cloth has been placed under this billet air conditioning compressor to prevent scratches and paint nicks on it and the valve cover under it. Large cloths, towels, or blankets work well to protect fender aprons and firewalls while installing engines.

cost you a lot of money and can easily make the difference between a good paint job and a great one.

Detecting Hidden Flaws

An automobile's engine compartment is home to a lot of different assemblies, components, parts, and systems. Along with an engine, the compartment houses brake system assemblies, heating and air conditioning parts, windshield washer lines, steering controls, suspension pieces, frame structures, engine cooling mechanisms and a host of other things. When restoring such a vast array of parts, one must look carefully to find hidden flaws that could escalate into major problems down the road.

While an engine and its accessories are out of an engine compartment, restorers have a perfect opportunity to check for structural problems. Access to remote engine compartment areas are unobstructed and with paint removed, bare metal can be thoroughly inspected. Look into cracks and crevices for early signs of rust or corrosion. Pull parts off that could hide warning signs that cancerous rust is spreading. If need be, tow your vehicle to a frame and brake specialist to have frame and suspension members checked for stability.

Frame, brake, steering, and suspension systems are integral to a vehicle's operation. Therefore, these systems must be inspected or put together by qualified professionals, especially for hot rods and those vehicles scheduled for alteration. Granted, a suspension and steering system may not be able to undergo adjustment and alignment without the weight of an engine. But these assemblies can certainly be checked for metal fatigue, hairline cracks, and other problems that would be easier and less expensive to repair at this stage than after the engine compartment has been outfitted with new parts, paint, and detail work.

Flaws may also include corroded bolt mounts, broken seam sealer coatings, cracked welds, and dented inner fender panels. These items must be repaired before extensive paint work is started. Rather than just putting more seam sealer on a bad spot, choose to remove old seam sealer and apply an all-new coating. Do the job right so it will last a long time. With regard to the restoration and maintenance of his Jaguar, Hall admits that the majority of work involves a lot of elbow grease. Sure, a lot of money can be spent on new parts, but just how good are they going to look on a pristine car that is idled on the side of a road with a broken tie rod, cracked motor mount, or other suspension, steering, brake, or frame problem?

Restoration Work

Removing and dismantling an engine and engine compartment can be a lot of fun. But one can get carried away. Work progresses quickly and before one knows it, all that remains is boxes of parts and an empty engine compartment. The problem with this kind of energetic activity is the mistaken impression that progress is being made. In reality, haphazard and random removal of engine compartment parts creates more work than it actually accomplishes.

A better, more productive approach is to remove parts one by one, clean, service, and properly label them before putting them away in a box for storage. Besides allowing you to study how various parts are removed, cleaning parts as they come off a car enables the restorer to inspect them and make a list of new parts as needed. A detailed labeling system, complete with pertinent installation notes, will be most helpful a few weeks or months later when installation begins and critical nuts, bolts, screws, and fasteners are needed.

To maintain these short-term goals, a box or two of clean and related parts can be reconditioned, painted, and detailed in an afternoon, evening, or weekend. Soon enough, all of the engine compartment accessories will be ready for systematic installation, with required new parts on hand.

An empty engine compartment needs to be thoroughly cleaned before restorative work can begin. Remove pockets of crud found along frame members, suspension pieces, firewall, inner fenders, and so on. Be sure to plug fuel lines, automatic transmission cooling lines, firewall openings, and other exposed areas to prevent water or cleaning agents from seeping inside. The more an engine compartment is dismantled, the better the access will be to parts such as the front grille section, radiator supports, splash shields, and inner fender assemblies. Definitive cleaning should also include all areas exposed by dismantling.

Engine compartment body areas that show signs of rust must be sanded down to bare metal. Use a high-speed rotary sander and 36 or 40 grit sandpaper to completely remove paint, scale, and rust deposits until bright, shiny metal is exposed. In tight spots, you may have better luck using the edge of an abrasive disc designed for automotive paint removal. The discs operate off of power drills to 5000rpm maximum and can be found at autobody paint and supply stores and through mail-order outlets like The Eastwood Company. Bare metal showing signs of rust should also be prepped with a metal conditioner to remove all rust. Chemical metal conditioners etch metal in preparation for paint. For extra-heavy rust concentrations, place rags soaked in conditioner over metal and leave overnight. Rust accumulations should wipe off easily the next day, and, at most, leave behind only small pockets of rust that may require additional restorative attention.

Rust-through problems can be corrected in one of two ways: install a new body part, or weld in a patch panel. New sheet metal parts are widely available for most automobiles through auto dealerships and specialty outlets like Mustangs Unlimited, Incorporated, of Manchester, Connecticut, Sherman

& Associates, Incorporated, of Roseville, Michigan, and Auto Body Specialties, Incorporated, of Middlefield, Connecticut. Available body panels range from battery trays to fender aprons, and splash shields to radiator supports.

Small rust-through holes can be repaired by panel patching. Basically, all rust-infected metal is cut out of a section and a new piece of sheet metal is welded in its place. Patches can be cut out of old door skins, quarter panels, and the like. They are positioned behind holes and then welded in place from

A nineteen-piece buffing outfit. Tools like this are used to buff and polish stainless steel, brass, diecast aluminum, plastic, or pot metal parts to perfection. Before installing such parts in your restored engine compartment, take time to buff them up to a mirror-like finish. This way, they will be detailed from the start and you will not have to try and polish them later in the confines of the engine compartment. Photo courtesy The Eastwood Company

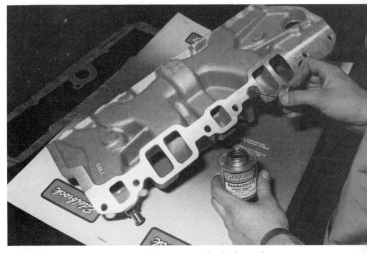

Engines must be put together correctly before they are reinstalled in the engine compartment. Be sure that all surfaces requiring gaskets are completely clean before applying gasket sealers and gaskets. Follow label instructions and recommendations for successful installation on the first try. Photo courtesy Edelbrock Corp.

Holding a power steering unit, alternator, or air conditioning compressor taut against its belt while tightening bolts is not always an easy task. To help maintain tension on parts, consider using a belt tightener tool. It fits between the belt and part and is then adjusted until the right belt tension is achieved. Place a cloth or towel between the end of the tool and engine part to prevent scratching. Tool courtesy The Eastwood Company

The paint on this bolt is a factory mark made by a worker to let other workers know it had been torqued and secured to factory specifications. Factory markings like this are important to concours competitors and those restoring automobiles to perfect factory originality. Photos taken before dismantling a restoration vehicle will serve as a reference for reapplying identical marks later.

the front. Welds are then ground smooth and areas are sealed with body filler or leading-in work.

Welding maneuvers on relatively thin sheet metal panels must be done with low amperage. The Eastwood Company offers a number of low-amperage welding units, as do tool houses and welding supply outfits. Because thinner metal tends to warp or burn through when exposed to prolonged high-heat conditions, welders must use low amperage and only weld short 1 to 1½in sections at a time. This is called stitch welding. Mycon prefers to use a wire-feed welder for this work. He immediately covers each weld with a sopping wet towel to absorb heat, cool panels, and shrink metal back to shape. Other welders like to flow cool, compressed air over new welds to dissipate heat and prevent panel warping problems.

Before welding or applying body filler material, metal must be perfectly clean and paint free. Dirt, grease, or paint residue could contaminate welds and weaken them. Unless you have welding experience, practice first on an old door, hood, or trunk lid. Be aware of potential fire hazards such as nearby cans of solvents or thinners, insulation material, brake lines and padding, or carpeting on the opposite sides of firewalls.

Part of a detailer's finishing touches includes adjusting and dressing rubber grommets and the wires or hoses that fit through them. The condition of this grommet and hood bumper on John Hall's Jaguar is excellent. He maintains them with dabs of dressing applied with a cotton swab. All excess is buffed off with a clean cloth.

As with welding, contamination will hinder the ability of body filler materials to adhere with maximum strength. Read and follow application instructions before putting filler on any panel. A variety of name-brand body filler materials are available at autobody paint and supply stores. Typically, bodywork technicians scoop out body filler with a putty knife and spread it out on a piece of metal or plexiglass. Hardener from a small tube is added per label instructions, and then mixed with filler until a consistent color is achieved. Material is then spread over a repair area with a plastic squeegee.

A hardener product makes filler material set quickly. Per instructions, allow plenty of time for all solvents to evaporate and filler material to harden before sanding. Use 40 grit sandpaper and a sanding block to initially knock down globs of filler and smooth surfaces to a basic finish. Follow up with 80 grit sandpaper to blend repairs with the surrounding surface. Rub your hand across repairs to feel for smoothness and conformity to surface features like lips, curves, and angles. Hand sensitivity can be increased by placing a clean cloth between your hand and the work surface.

Many meticulous bodywork technicians prefer to apply a thin coat of finish glazing putty over body filler repairs. The purpose for this is to cover fine sanding scratches and fill pinholes left behind after regular body filler surfaces have been sanded smooth. Body fillers strengthen repairs while glazing putty materials seal those areas in preparation for paint work. Check with your local autobody paint and supply store jobber as to which brand of glazing putty is recommended for the paint system you intend to use.

Even though an engine compartment body panel may be repaired with filler or hot metal work, it will not be ready for paint until the surface has been coated with primer materials. By all means, bare metal should be coated with an epoxy primer, like PPG's DP 40. Primer-surfacers are composed of solid materials that are sprayed over repair areas to cover fine sanding scratches and other surface blemishes. Once primer-surfacer coats have dried, 400 to 500 grit sandpaper is used to smooth surfaces to perfection.

Painting

Painting an engine block, accessory, or engine compartment body panel requires the same attention to detail as painting a car body. Masking must be precise, paint booth conditions clean and free of any dust concentrations, surface preparations meticulous, and materials mixed according to label instructions. Plenty of technical assistance is available through autobody paint and supply stores. Jobbers can answer questions and help with product purchases. In addition, information sheets and application guidelines from paint manufacturers are available at these stores on any product you choose to use.

Depending upon the type of paint system you have chosen to apply, a sealer material may be needed. Sealers are products used to prevent actual paint ingredients from being absorbed by primer materials. Autobody paint jobbers can recommend the type you'll need, or you can consult paint product information sheets displayed at the store. Sealers do not generally require any sanding. Adhesion promoters work much the same way as sealers. Their function is to provide an excellent base for the adhesion of new paint materials. Again, use of these products is determined by the type of paint system you employ.

If you have chosen to paint everything inside the engine compartment the same color as its vehicle body, you may have to paint some plastic or flexible parts. For those, you must include prescribed amounts of a flexible additive with each paint application. Without a flexible additive, paint will crack, peel, or flake after parts have flexed during normal vehicle operations. The exact mixing ratios are provided on labels and in application guideline sheets available at autobody paint and supply stores.

If your engine compartment had been subjected to repeated multipurpose dressing applications on hoses, wires, plastic, and rubber parts, chances are surrounding metal body panels are saturated with silicone dressing residue. In extreme cases, silicone has been known to penetrate paint layers and impregnate the metal surfaces. In that situation, the paint will require a fisheye eliminator additive. Without it, silicone-based materials on metal surfaces will cause paint to fisheye, which is evidenced by small blemishes on paint surfaces that look like micro volcano eruptions. Every auto paint manufacturer offers its own brand of fisheye eliminator, such as PPG's DX 77 Fisheye Preventer, Glasurit's Antisilicone Additive, and DuPont's Cronar Fish Eye Eliminator 9259S.

Body surfaces must be perfectly clean before applying any paint material. Mycon uses air pressure to blow away the major accumulations. Then he dabs one clean cloth with a wax and grease remover, wipes it across a surface, and follows it with a clean, dry cloth to pick up moisture residue and lingering contaminants. Going a step further to ensure perfectly clean painting surfaces, he sprays on an aerosol glass cleaner and wipes it off with another clean, dry cloth. This helps to remove moisture and pick up small traces of dirt or debris.

Tack cloths are specially made to pick up lint and dust from surfaces just before painting. They are sticky and work well to attract particles that could otherwise create nibs on paint finishes. Auto painters unfold tack cloths and use them in a loose wad, as opposed to folded flat. This final task should be completed moments before spraying paint to ensure dust-free surfaces.

In decades past, auto painting was limited to either enamel or lacquer materials. Today, a wide range of paint products are available. More than ever before, auto painting has become a high-tech business. Painters must be aware of special additive usages, thinner and reducer temperature ranges, and a host of other technical data. An autobody paint and supply jobber can help novice painters select a paint system most appropriate for their needs.

Along with tens of thousands of paint colors to choose from, restorers must evaluate which paint system will work best for their projects. Custom paint jobs featuring pearl or candy finishes may work great for show cars and those seldom driven because nicks and scratches will be held to a minimum. However, touching up such custom paint finishes is often impossible. Conversely, everyday drivers might be best off with a durable catalyzed urethane paint that can be sanded early for the removal of imperfections, will hold up well in most conditions, and can be easily touched up without blending problems.

While contemplating your engine compartment's paint job, do not overlook the need to mask. Nothing, it seems, looks worse than paint overspray on parts not designed to be painted. Use only automotive-quality masking tape and masking paper designed for painting use. Newspaper tends to let paint bleed through, and generic masking tape will frequently leave behind glue residue along with irregularly sealed edges that cause paint blemishes.

Engine Compartment Assembly

Regardless of an automobile's intended function, its restored engine compartment should be put back together with the same care and attention afforded one destined for national concours championships. Make use of automotive masking tape and plenty of soft cloths or towels to guard against accidental bumps and scrapes as the engine and other parts are lowered in.

If you heeded earlier advice about cleaning parts as they were extracted from your car's engine compartment, their installation should be relatively quick. However, if that advice went unheeded, you will now be faced with cleaning and reconditioning those parts that need it. Once the engine is in, parts

These tools are used to straighten radiator and air conditioner condenser fins. The Fin Rake is a six-sided tool with slots designed to fit an assortment of condenser patterns. It works well for straightening fins into a

uniform pattern. The Radiator Fin Pliers features wide blades that are thin enough to easily fit between radiator fins and then grab hold to straighten them out. Tools courtesy The Eastwood Company

144

cleaning and detail work should start with those parts scheduled to be installed first, like the radiator, power steering unit, alternator or generator, and so forth. Once they are cleaned, painted, and detailed, you can install them and watch your car's engine compartment start to come back to life.

It is very important that the engine and engine compartment be put back together properly. Failing to install what you may deem trivial items could adversely affect an engine's overall ability to perform. Such is the case with thermostats on engines equipped with onboard computers. An engine missing a thermostat could run so cool, the computer could be fooled and will keep the fuel and ignition system locked in the warm-up mode, a condition that will cause the engine to run way too rich.

Smaller accomplishments, like installing everything on the front of an engine that belongs there, are very rewarding and should inspire you to keep going. As mentioned, it helps to detail parts before installing them. Hoses, for example, are much easier to clean and buff while off of an engine. The same goes for painting housings on alternators and generators, or detailing fender apron-mounted parts like battery trays, windshield washer containers, voltage regulators, heater hoses, and the like.

Bolt and screw heads are best painted while screwed into a piece of cardboard. Threads are protected and there is plenty of room for maneuvering. Wrenches or sockets will, of course, pose scratch hazards during their installation and tightening. So, in an effort to avoid such blemishes, try placing a piece of plastic over bolts or screws to cushion and guard against metal-on-metal contact while torquing them down. Polished pieces should be handled with a soft cloth. Hold a chrome or billet bracket, for example, with a cloth draped across your hand. This will prevent skin oils from smudging the part's surface.

While installing hose clamps and other outside connectors, try to line them up so that they all point toward the same direction. This is a minor detail, but makes for a more uniform-looking engine compartment. Arrange electrical connectors in the same manner. Balance and uniformity contribute to an engine's pleasing appearance and are proof that you paid attention to the details.

Finishing Touches

After laboring for months, weeks, or days over the engine compartment of your favorite vehicle, do not stop short of attending to the details. Sure, the engine compartment body panels, engine, and accessories look great, but what about the little things that need attention?

Many engine compartments are adorned with decals, stickers, and emblems of all kinds. These minor items help to bring out a vehicle's personality. Classic car concours judges are keenly aware of what is supposed to be in an engine compartment and will quickly deduct points for a missing air cleaner sticker or valve cover decal denoting the cubic inch capacity of its engine. When installing decals, stickers, or emblems, never put your fingers on the backside (sticky side). You could soil the piece, or your fingerprints could become permanently affixed.

What about the black paint that is supposed to adorn grooves on certain valve covers, or the alignment of hose clamps that could so easily balance out the engine compartment? How about that paint chip on the fender apron that occurred during engine installation? Or the scratch on the radiator side piece that occurred during its insertion?

Rubber grommets should fit snugly, wires must be routed tightly and symmetrical with the engine part they rest upon, belts have to be tight, and battery

The difference between a concours winner and a consistent fourth place finisher is generally in relation to the amount of time each participant spends taking care of small details. Hall spends a lot of time polishing and cleaning his prized 1959 Jaguar XK 150 S. That time and effort is rewarded with first place concours awards. Follow Hall's example and pay close attention to the details when finishing up the last few items of your complete engine compartment restoration to make it sparkle and shine like a blue ribbon concours champion.

connections should be shiny. In your zeal to put an engine compartment back together, do not forget all of the little things that need attention. Engine detailing is a meticulous job. No matter what type of vehicle you are restoring, take the few minutes or hours needed to make it pristine. Tasks may only involve tightening a set of ignition wire holders or using an artist's paintbrush and engine paint to touch up paint nicks. But find the time to do it.

We have all seen novice car enthusiasts show off an engine compartment full of brightly painted parts, some of which were supposed to be painted and some that were not. It is this time-consuming attention to detail that separates the novice from the experienced detailer. Lacquer thinner will remove paint overspray from just about anything, and it's the detailer who wants to do the best job possible that spends extra time correcting those oversights. It's usually the same person who finishes first in car shows, concours, and other auto events.

Chapter 11

Car Shows and Storage

Car shows of all types are held throughout the country on a regular basis. Spring and summer months are most popular, especially in regions that experience winters with lots of rain or snow; few car enthusiasts relish the thought of driving their special automobiles over wet or slushy roads. Those who can afford it haul show cars to events inside enclosed trailers, or hire an auto transport company to deliver cars by way of an enclosed trailer, thus avoiding any road mishaps that could result in paint nicks, scratches, or road grime accumulations. Once their automobile is parked in its position at a show, entrants frequently spend time dusting off engine compartments with window cleaners, wiping up minute spots of oil, and making sure every square inch of their car or truck is spotless and shiny.

Putting a special automobile together requires a lot of time, effort, and financial commitment. When detail efforts prove successful and project cars are finished with terrific results, owners generally like to make sure everyone gets a chance to see the fruits of their labor. Car shows offer enthusiasts this kind of opportunity, along with lots of other interesting things to see and do. During the car show season, owners are able to start up and drive their cars once in awhile and also spend time cleaning, polishing, and servicing various parts and accessories.

During the winter, many special automobiles are put into storage. Some owners of very rare or expensive vintage and classic models store them in custom-made garages equipped with humidity, heat, and ventilation controls. Strict maintenance of the

Trailers hauling automobiles are common sights around concours events and car shows. Many of the special show automobiles are much too nice to be driven on regular roads. John Hall's trailer is equipped with lights and a large rack for storing detail supplies and his library of Jaguar literature.

atmosphere and environment inside these garages goes a long way toward helping to preserve car bodies, upholstery, tires, and other vehicle components. Automobiles must be properly prepared for storage, however, to reduce chances of parts becoming frozen in position, faded by sunlight, rotted through constant moisture contact, or ruined in other ways.

Winter weather conditions may hinder an enthusiast's ability to keep his or her car in top condition, and the lack of car show events might find cars sitting idle for extended periods, maybe months at a time. It is just before these off-season spells that cars should be properly serviced and prepared for storage. Automobiles that have been stored correctly should make it through the winter with little difficulty and ultimately require only a few detailing and servicing chores at the beginning of the spring car show season.

Car Shows

All kinds of organizations sponsor car shows. Most are put on by local car clubs, but many are organized by nonprofit groups in hopes of making money by charging patrons a nominal entrance fee. Some are set up by professional car-show companies for the enjoyment of all automobile enthusiasts. Specialty car auctions are frequently coordinated with car shows, auto swap meets, and flea markets to attract larger crowds. Car clubs with special interests in certain vehicle makes and models will host shows specifically for those kinds of cars, like Thunderbirds, Mustangs, 1955-57 Chevrolets, and so on.

You can learn about local, regional, and national car show events through automotive magazines and local automotive-related businesses such as parts stores, high-performance shops, and the like. *Hemmings Motor News* carries hundreds of auto-related events in each issue. Most listings are free to advertisers as long as sponsors meet certain guidelines. Events are generally advertised a couple of months ahead of time. Ads include addresses and phone numbers of people to contact for information about participation requirements. Many other auto magazines also list calendars of events and most of them frequently offer this service free to sponsors who fit within specified guidelines.

Local car shows are frequently put on just for the enjoyment of local auto enthusiasts. Car clubs get together and plan events at local parks, college campuses, or other open sites. This gives members a chance to display their favorite cars and also get together to exchange ideas and talk about their cars. These meets are perfect for novice auto enthusiasts; they can talk to car owners about engine compartment restoration techniques, parts availability, paint colors, and more.

Automobile trailers must have a means to hold vehicles securely in place. A set of floor anchors and a strap equipped with a come-along is used to keep Hall's car anchored. Electrical outlets are connected to a shore power inlet so that lights can be used in the trailer for detailing purposes.

A large shelf at the front of the trailer is used to store items like detailing kits, cleaning cloths, auto literature, and spare parts. Note the carpet material that covers the trailer's fender indentations. This was installed to prevent accidental paint chips while opening car doors inside the trailer.

Larger car show events are usually held over an entire weekend. Owners are allowed to bring their cars into an auditorium or building site on Thursday night or anytime on Friday. This gives them a chance to wipe off tires, polish as needed, and lightly detail engine compartments in preparation for the show's opening on Saturday morning. Signs are posted next to vehicles listing the names of individuals or shops that did restoration work on them, such as painters, upholstery shops, high-performance mechanics, and so on. Many participants dress up their display by featuring an array of memorabilia relating to their model year or era. Some displays are quite creative and interesting.

Car shows often present awards to vehicle owners for various accomplishments. For instance, there's the award for best-looking engine compartment, best upholstery, best paint job, and most popular display, to name just a few. Sometimes visitors are asked to vote for their favorites, other times only participating entrants vote, and other events combine ballots from both visitors and participants. Whichever the case, winners walk away with nice looking trophies or awards which may consist of car parts or other auto-related items.

The amount of time allotted for last-minute cleanup and detailing before the start of a car show or

Car show participants can win awards and prizes for a multitude of things, from best upholstery to who traveled the farthest to get to the show site. John Pfanstiehl's 1959 Cadillac Eldorado should win prizes for having the lowest mileage of any 1959 car in existence, and also for being so original. Cadillac used old tires as muffler hangers, evidenced by a strip of whitewall tire on the right side of the muffler. This should surely be a concours point-winning item.

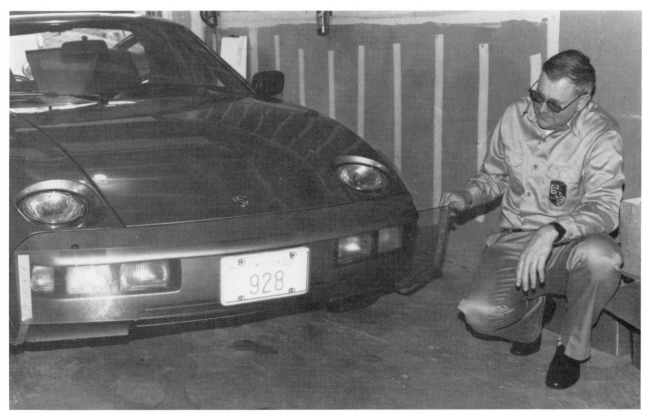

Ridderbusch built this clear plastic shield for the front of his 928 before a trip to a national Porsche concours in Washington, D.C., a few years ago. It did a great job of protecting the front bumper and hood area from paint chips and nicks, while still allowing plenty of headlight illumination through it.

concours event varies with each occasion. National concours competitions may allow three days' preparation time, while local or regional events allow only a few hours. Ridderbusch drove his 928 to Washington, D.C., one year for a national Porsche concours. Although the car was cleaned and detailed prior to the trip and self-sticking shelf paper was applied over the bottom of the car, he and his wife spent three days cleaning it at the concours site before judging. Hall brings his XK 150 S to events in an enclosed trailer, but still spends time touching up hoses, polishing paint, and dressing tires after his arrival.

When judging is complete, Hall frequently drives his car in featured slalom competitions or other driving events. These fun experiences always result in his car getting dirty from chalk lines and other road debris. More often than not, Hall will put in more than twenty hours cleaning the engine compartment and undercarriage after driving in a slalom or rally. The time and elbow grease he puts into cleaning are well worth it, however; he totally enjoys driving the car, and his family has a great time riding in it.

Car Show Preparations

Almost everyone who displays a car or truck at a car show, and every competitor at concours, brings along a detail kit. Enthusiasts carry small bottles of touchup paint, thinner, artist's paintbrushes, polish, dressing, and assorted soft cleaning cloths. Hall uses old baby diapers that are sold by diaper service companies as rags. They are very soft and have never been subjected to hard debris such as metal filings, grit, and so on. He has noticed, however, that once in a while an old diaper will wad up between cloth layers, which could be a scratch hazard. So before using old diapers, feel the material to be sure no such problems exist.

Engine compartments on vehicles that have been driven to shows or events will most likely be dusty and may have a very thin oil-like film on some

For the long trip back to Washington, D.C., Ridderbusch applied self-sticking shelf paper to the undercarriage and also to large exterior mirrors before installing a cover on them. The car's excellent wax protection prevented the shelf paper from adhering too strongly, and the paper prevented paint scratches from covers as they vibrated in the wind.

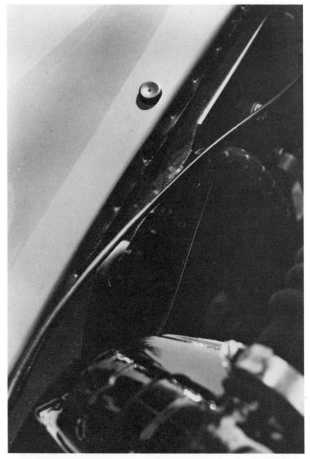

A slight hint of dust is seen on the fender apron inside the engine compartment of Hall's Jaguar—a result of his participation in a slalom. It will have to be completely removed from all parts of the engine compartment before the next concours, a task that will easily take twenty hours. Now, that is really paying attention to detail!

parts. A cloth dampened with water can be used to wipe off dust, and dabs of Simple Green on a cloth works well to remove greasy films or oil spots. Plan to polish chrome and other engine brightwork, and remember to check radiator fins for bugs and debris. Hall uses a thick, wide artist's paintbrush to dust off engine compartment areas too tight for cloths or towels. Long cotton swabs are used to pick up oil from around gaskets, bolt heads, and assorted grooves or crevices.

Polish residue normally turns white when it dries. This is broken loose and whisked away with a cutoff paintbrush. For last-minute touchup on thin gasket edges blemished with hints of dry polish, Hall applies Meguiar's #40 with a cotton swab. He says this makes gaskets look clean and covers dried polish just long enough to endure judging. Be sure to look under hoses and wires, along seams, and around hood hinges for dried polish.

Small baby food jars or other glass containers with a tight-sealing cap are great for holding small amounts of touchup paint. Your detail kit should include jars with each paint color used in your vehicle's engine compartment. Include a jar of thin-

ner or reducer, too, for cleaning brushes. As you inspect your engine compartment before showtime or judging, use these materials to touch up small nicks or other blemishes. A black felt marker works well for emergency touchup on black parts, belts, and some hoses. Bright silver paint is effective as a temporary last-minute touchup on chrome nicks, bolt heads, some hose clamps, and similar parts.

Under the engine, look for dust, road grime, oil spots, globs of grease on front suspension and steering parts, antifreeze stains near radiator overflow tubes, and dirt or grit accumulations on inner fender ledges. A drop light is handy for this job, as are a creeper, plenty of cleaning cloths, a roll of paper towels, and maybe a pair of clean, lightweight coveralls.

A bucket of clear water and a cotton wash mitt may work best for wiping off areas under the engine, especially for those cars that were driven or towed to

This detail kit contains a set of artist's paintbrushes, masking tape, cotton swabs on long, wooden handles, vials of touchup paint and thinner, and a few other goodies. Car show participants should carry similar kits with them to take care of last-minute detailing tasks before the opening of a show or start of judging.

One of the last tasks Hall likes to complete before concours judging is light dusting of engine compartment areas with a soft artist's paintbrush. This removes traces of lint from cleaning cloths and dust that may have settled while the vehicle was parked. There may not be much dust to remove, but to him, this effort could easily be worth a few tenths of a point.

151

events. A spray bottle of Simple Green should be kept handy to mist cloths that are needed for cleaning spots too stubborn for just plain water. After wiping parts with a damp cloth, be sure to dry surfaces with a dry one to remove moisture streaks and any traces of lingering dust or dirt residue.

Owners of show cars that have beautifully detailed undercarriages and sparkling lower engine compartments, frequently prop large mirrors under them so admirers can easily appreciate their beauty. Small light fixtures can be set up under cars to help highlight chrome pieces and perfectly maintained assemblies. Some owners have gone to great lengths in adorning this part of their car or truck with special pinstriping and other unique paint schemes.

The free time you have before the official opening of a car show or other auto-related event is valuable. You can accomplish a few last-minute detailing tasks and also have opportunities to exchange ideas with other show car owners, talk to vendors who sell auto parts or other auto-related

A favorite emergency measure among car show enthusiasts is the use of a black felt tip marker to touch up minor scratches or blemishes on black parts. This works great as a temporary measure, and novices should include one in their car show detail kit.

materials, and learn a few tips from restorers and other professionals who may have a display to advertise their services.

Because there are so many different things to do at a car show, your vehicle should be clean and detailed before arriving. This will give you more time to wander around and talk to people instead of having to spend it all on preparing your car. Novices and serious enthusiasts alike are encouraged to take lots of pictures of engine compartments set up the way they like them. Ask those car owners how they accomplished specific detailing or restoration tasks and even write notes, if need be, on how jobs were done or where certain parts or services are available.

Preparing Engine Compartments for Storage

Ridderbusch believes the best way to keep engines in top condition is to run them at least once a month. Operating them for a while at normal operating temperature helps to evaporate accumulated moisture from exhaust systems, and lubricate all internal engine parts. Engine operation exercises moving parts and keeps pliable seals from becoming brittle. Driving cars also exercises brakes, steering, suspension, and other assemblies to keep them lubricated and in working order.

Starting and driving special cars is not always a viable option, however. Out-of-town trips, inclement weather, and other factors frequently prevent owners from being able to do this. For them, long-term idle storage is unavoidable. To ensure that your automobile is protected from damage while in storage, a number of precautions must be taken.

An ideal environment for long-term storage would include a separate garage with controlled humidity, ventilation, and heat. It would have no windows or skylights to let sunshine in, and would include a means to keep all rodents and insects out. A regular garage is another option, followed by a carport or tent structure. Auto magazines often carry advertisements for temporary tent-like enclosures for automobiles that offer protection against inclement weather and outdoor dirt, tree, and airborne debris.

The entire engine compartment should be thoroughly cleaned before it is put into storage. Elimination of grease and dirt helps to minimize absorption and containment of moisture. After washing, drive your vehicle to blow off lingering pockets of water and to heat up all internal engine fluids. Change the oil and filter while the engine is still warm. This is so particles of sludge and other contaminants picked up and suspended in oil while the engine is operating can be drained along with the oil. Then put on a new filter and fill the crankcase with fresh oil.

Drain and flush the cooling system completely, including the heater's radiator. Refill it with a 50:50 mixture of antifreeze and water. Some serious enthusiasts believe distilled water is best because it contains no foreign materials or additives. You might also consider putting in a rust inhibitor and a water pump lubricant, both available at auto parts stores.

All other engine compartment fluids should be topped off. Consider changing power steering and brake fluids as well. Under normal circumstances, power steering fluid should be changed every 40,000 miles and brake fluid every 25,000 miles. Under normal operations these fluids will pick up contaminants that should be eliminated before storage.

Once all fluids have been changed, drive your car or truck to a busy gas station and fill up the fuel tank with a quality gasoline. An active gas station is best because it is more likely to have fresh fuel not polluted with moisture condensation or other storage tank debris that can be a problem. A full fuel tank reduces the amount of space needed for moisture-laden air to begin a rusting process on the exposed fuel tank walls. Add a can of gasoline stabilizer to help reduce chances of fuel deterioration. While driving your car home, be sure to run your car long enough to thoroughly mix gasoline stabilizer into the fuel so that it has an opportunity to flow through fuel lines and into the carburetor. This will help reduce gum and varnish build-up inside those units.

With your automobile parked in its storage spot, pull the spark plugs and pour a tablespoon of clean oil into spark plug holes. Turn the engine over a few times with the plugs out to coat cylinder walls with the fresh oil. Coat spark plug threads with an antiseize product before they are put back into the head.

Disconnect the battery and take it out of the vehicle. Be sure it is filled to the proper level with distilled water and its case is clean. Clean battery terminals and coat them with dielectric grease. Store batteries in a clean, dry place, preferably on top of a piece of wood or rubber to minimize current loss. Clean and brighten battery cable ends and coat them with dielectric grease. In fact, it is a good idea to clean and then coat all exposed electrical connections with dielectric grease to prevent corrosion during storage.

Next, stuff a clean cloth into the tailpipe and seal it with a piece of thick plastic held in place with string or heavy-duty rubber bands. This will prevent moisture condensation in the exhaust system which could also move up through open valves into the heads and cylinders. This evolution also eliminates a place for rodents to nest. Gently fill carburetor throats with a clean cloth to prevent moisture from entering the intake manifold. Seal off air cleaner intakes with a cloth and plastic like you did on the tailpipe to keep moisture and rodents out. Wrap up baseball-sized packets of camphor (moth) balls in cheesecloth and place them around the engine compartment to keep insects and rodents away.

Be certain all paint chips are touched up and a fresh coat of wax is put on all painted surfaces. This will do a lot to prevent moisture from adversely

affecting those surfaces. Unpainted, bare metal pieces may be lightly coated with a penetrating oil to prevent rust accumulations; such parts might include exhaust manifolds and header pipes. A drop of lubricating oil may also be needed in the cups of exposed and unsealed generator and starter bearings on older cars, as well as in the grease cups on antique auto distributors.

Cars put into storage should be covered with a quality car cover made of a material that breathes. Plastic tarps are not good because they permit condensation under them with no way for it to escape. A cover will also prevent sunlight from fading interior upholstery or paint. Every couple of months, or so, plan to turn the engine over a few times to lubricate internal parts and exercise valve springs. Be sure the parking brake is left off, so brake shoes do not become frozen to their drums.

When a vehicle is taken out of storage, let the engine warm up slowly and do not rev it until it has

Automobiles should never be put into storage in a dirty condition. Dirt and grease accumulations attract, absorb, and hold moisture that will eventually lead to rust and corrosion problems. If you want your special car or truck to survive storage in excellent condition, plan to detail the engine compartment thoroughly before parking it in its storage spot.

warmed up to normal operating temperature. This should ensure all parts have been fully lubricated and exercised before undue strain is put on them. Operate all accessories such as heater, lights, defroster, windshield wipers, brakes, and so on. Because your car or truck has been in storage for some time, it is not known if corrosion or other problems cropped up that would render these features inoperable. The first few driving miles must be slow ones. This gives you a chance to check the operation of all the various components, along with keeping an eye on the oil pressure, water temperature, and electrical gauges.

Overview

A detailed engine compartment will not stay clean and shiny all by itself. Normal engine operation will coat motor spaces with oily films that will eventually attract dirt, dust, and grime. Battery terminal corrosion can seep down inner fender walls and corrode the sheet metal, and slight radiator coolant overflow will stain whatever it touches.

The best way to maintain a good-looking engine compartment is to clean it on a continual basis. Although a full-blown detail does not have to be done every month, you would be wise to wipe off the underside of the hood, fender aprons, radiator, air cleaner, and other surfaces every month, or so, with a little Simple Green or other cleaner on a cloth. As paint starts to dull, polish it with a mild glaze and then protect it with a quality carnauba wax.

Hoses and wires do not require a lot of dressing to maintain a high-luster appearance. Simple clean-

ing usually keeps them looking good. Too much dressing can sometimes attract dust and dirt, so be sure to wipe off excess after every application.

While you are cleaning parts of your engine compartment, check the tightness of bolts, screws, and other fasteners. Valve cover hold-downs are notorious for coming loose and allowing oil to seep past gaskets. Prevent this kind of mess by periodically tightening them. Other normal engine preventive maintenance chores are also easier and less of a hassle when engine compartments are clean.

In-depth engine and engine compartment detailing can be fun and rewarding when it is approached with realistic expectations. A quick power washing and light scrubbing will improve many engine compartments, but those efforts will be short-lived.

Hoses do not need a heavy dressing application before storage. They do need to be clean and free of oily deposits, though. This is the factory-stock radiator hose that came on Pfanstiehl's 1959 Cadillac. Since the car has been driven less than 2,300 miles, it should be safe to say that this hose has experienced quite a few years of storage. It still exhibits a factory yellow line down its left side, and looks good overall for not having received dressing applications. Cleanliness is the key in preserving automotive engine compartment parts and accessories.

Paint chips and nicks in the engine compartment must be touched up before storage. This ChipKit may be just the thing to help you accomplish that task. Paint chips that expose bare metal must be covered before storage to prevent rust from getting a head start and possibly spreading throughout parts or panels. Photo courtesy Pro Motorcar Products, Inc.

Accumulations of grease and oil attract more dust, dirt, and oily deposits. A haphazard attempt at engine cleaning will surely result in missed pockets of grease and grime that will rapidly increase in size until, in seemingly no time, the engine and surrounding compartment will look just as bad as it did when you washed it the first time. Along with that, hurried attempts at painting blocks and other engine parts will almost always look as if only a small amount of effort was put into the work, and offer no more than a flavor of amateurish detailing.

Expect to spend an entire day cleaning the engine compartment of your car. Plan another full day for attending to paint and polish details. As you become more aware of how well your work efforts are paying off, don't be surprised if you end up taking three or four days to complete a meticulous and very satisfying quality engine compartment detail. Take your time, enjoy yourself, and good luck on your automotive ambitions, whether they be for simple self-rewards, car show trophies, or Best of Show Concours d'Elegance awards.

Complete Engine Compartment Detail

This section features seven photographs of Terry Skiple's 1974 DeTomaso Pantera receiving an engine and engine compartment detail, from start to finish. It includes a "before" photo, washing, cleaning, painting, and an "after" photo.

Terry Skiple's 1974 DeTomaso Pantera had been stored in a garage for a few years. It was not driven until recently, after a few minor mechanical problems were repaired, evidenced by the handprints and smudges on the engine compartment aprons. The engine appears as though no paint has ever been sprayed on it, and is covered with a rusty film. The valve covers are new as part of recent repairs.

Art Wentworth uses a small pressure washer to remove dirt and debris from the deck lid underside. The engine was covered with towels to prevent excessive water accumulations. Simple Green, a wash mitt, and paint-brush were used for cleaning. This section will be thoroughly washed and dried before detailing is continued.

Towels have been placed on top of the quarterpanel sections to prevent inadvertent scratches while leaning over them to clean engine compartment areas. A clean cloth was placed inside the carburetor throat and a plastic freezer bag taped over the carburetor to prevent water from entering that space. The distributor was also covered with a cloth and a sheet of aluminum foil. The engine was coated with degreaser and is now being thoroughly rinsed with a pressure washer.

In-depth cleaning on the engine and around the engine compartment was accomplished with Simple Green, a wash mitt, and assorted soft brushes. Here, a toothbrush is used to remove grease from around a valve cover gasket area. A combination of pressure washer power and degreaser removed the rusty film that covered the engine, revealing a faded coat of Ford Blue engine paint.

Parts of the engine have been covered with masking tape, and a few sheets of folded newspaper are used as a paint block while high-temperature engine paint is sprayed on the block and intake manifold. Masking tape covers the edge of new valve cover gaskets, an edge of the transaxle where it butts next to the block, hose ends and clamps, and other assorted parts not scheduled for paint.

Paint overspray was removed from parts with lacquer thinner. The bare transaxle housing was scrubbed with cleanser and #0000 steel wool, as nothing else would remove the encrusted film of dirt, grease, and stains. Here, Wentworth applies a coat of wax to the air cleaner lid after polishing it with a chrome polish. Other chores were accomplished too, like cleaning and dressing wires and hoses, polishing other items inside the compartment space, and painting items that suffered scratches or other blemishes.

The engine compartment in Skiple's Pantera after detail work was completed. Because of easy access to the engine and compartment space around it, detailing was relatively unobstructed. The entire project took about six hours. This is what engine detailing is all about!

158

Sources

Antique Automotive
4124 Poplar Street
San Diego, CA 92105

Auto Body Specialties, Inc.
PO Box 455, Route 66
Middlefield, CT 06455

B&M Automotive Products
9152 Independence Avenue
Chatsworth, CA 91311

Keith Black Systems
5630 Imperial Highway
South Gate, CA 90280

Corvette & High Performance, Inc.
2840-D Black Lake Boulevard SW
Olympia, WA 98502

Drake Restoration Supplies
4504 "C" Del Amo Boulevard
Torrance, CA 90503

Earl's Performance Products
189 W. Victoria Street
Long Beach, CA 90805

The Eastwood Company
580 Lancaster Avenue
Box 296
Malvern, PA 19355

Edelbrock Corporation
2700 California Street
Torrance, CA 90509-2936

Flex-a-lite Consolidated
4540 S. Adams
PO Box 9037
Tacoma, WA 98409

Gary's Plastic Chrome Plating, Inc.
39312 Dillingham
Westland, MI 48185

Harnesses Unlimited
PO Box 435
Wayne, PA 19087

Harwood Industries, Inc.
Route 3, Box 933 A
Tyler, TX 75705

Hemmings Motor News
PO Box 1108
Bennington, VT 05201

Jaguar Clubs of North America
555 MacArthur Boulevard
Mahwah, NJ 07430-2327

Kanter Auto Products
76 Monroe Street
Boonton, NJ 07005

Meguiar's, Inc.
17991 Mitchell South
Irvine, CA 92714

Metro Moulded Parts, Inc.
11610 Jay Street
PO Box 33130
Minneapolis, MN 55433

Mustangs Unlimited, Inc.
185 Adams Street
Manchester, CT 06040

Newlook Autobody
13205 NE 124th Street
Kirkland, WA 98034

M. M. Newman Corporation
24 Tioga Way
PO Box 615
Marblehead, MA 01945

Jim Osborn Reproductions, Inc.
101 Ridgecrest Drive
Lawrenceville, GA 30245

Pro Motorcar Products, Inc.
22025 US Highway 19 N
Clearwater, FL 34625

RB's Obsolete Automotive, Inc.
18421 Highway 99
Lynnwood, WA 98037

Scotts Manufacturing Co.
27833 Avenue Hopkins, #3
Valencia, CA 91355

Sherman & Associates, Inc.
28460 Groesbeck
Roseville, MI 48066

Joe Smith Automotive, Inc.
3070 Briarcliff Road NE
Atlanta, GA 30329

Spectre Industries
922 N. Ninth Street
San Jose, CA 95112

Super Shops, Inc.
4028 196th SW
Lynnwood, WA 98036

Taylor Cable Products, Inc.
301 Highgrove Road
Grandview, MO 64030

Thermo-Tec
PO Box 946
Berea, OH 44017

XKs Unlimited
850 Fiero Lane
San Luis Obispo, CA 93401

Index